# THE SCHOOL OF FENCING

*Domenico Angelo about 1760*
*Drawn by John Gwynn, RA, engraved by W. W. Ryland*

Domenico Angelo

# THE SCHOOL OF FENCING

## With a General Explanation of the Principal Attitudes and Positions Peculiar to the Art

Edited and presented by
Jared Kirby

Notes by
Maestro Jeannette Acosta-Martinez

Frontline Books

*The School of Fencing*
*With a General Explanation of the Principal Attitudes and Positions Peculiar to the Art*

A Greenhill Book

First published as *L'Ecole des armes* in 1763. The first solely English edition, which this edition follows, was published in 1787.

Published in 2005 by Greenhill Books, Lionel Leventhal Limited
www.greenhillbooks.com

This edition published in 2017 and reprinted in this format in 2024 by

Frontline Books
an imprint of Pen & Sword Books Ltd,
47 Church Street, Barnsley, S. Yorkshire, S70 2AS
For more information on our books, please visit
www.frontline-books.com, email info@frontline-books.com
or write to us at the above address.



ISBN: 978-1-39907-881-8



CIP data records for this title are available from the British Library

Designed by Ian Hughes

Printed and bound by CPI Group (UK) Ltd, Croydon, CR0 4YY

# Contents

# FOREWORD

IT'S HARD TO believe that over a decade has passed since the publication of this book. In that time small-sword has seen an unprecedented resurgence. More and more people are gravitating towards the great works of 18th-century masters such as Angelo to create the foundation for studying French small-sword.

This foundation is made easier by the clear, concise instructions provided by Angelo. His exercises are invaluable and his expression of the artistry required for this weapon remains unparalleled in the canon of French small-sword treatises. I continue to understand why Domenico Angelo was chosen as the best example of French fencing by the compilers of *Encyclopédie*. They introduced the fencing section (taken completely from Angelo's book) by saying, 'Had we known of any work more perfect in its kind, we should have made use of it.' His legacy continued for over a century after his death and here we are hundreds of years later still pouring over his work.

I cannot count the number of times people have thanked me for bringing this work back into mass publication. Those accolades go to Maestro Jeannette Acosta-Martinez for the annotations in this book as well as the appendices, which facilitate a deeper understanding of French small-sword. Without her expertise and passion for this weapon I highly doubt we would have seen this resurgence of interest. It has made it much easier to start studying this weapon.

My advice to those beginning to learn French small-sword is to follow closely in the steps of the Masters. Truth may be exhibited in two ways, word and deed. As this treatise is exceptionally valuable in its words, it is of greater merit to put into practice these words through the wisdom and guidance of a seasoned professional such as Maestro Jeannette Acosta-Martinez. This traditional approach to learning will expedite your studies and allow for a deeper understanding of fencing. With the beautiful symmetry seen between classical ballet, classical dressage and classical fencing, it would appear that both direct instruction from a qualified teacher and the support of the written word will be a valuable means of learning. In this way can we honour this great master and the treatise left to us.

Jared Kirby, Maestro d'Armi 2016

# INTRODUCTION

BY JARED KIRBY

DOMENICO ANGELO TREMAMONDO was one of the most famous swordsmen of the eighteenth century. His treatise *L'Ecole des armes* (*The School of Fencing*) was the most popular book on fencing for more than half a century. Though his fame spread throughout Europe, Angelo was especially renowned in England and France, and to this day he is remembered as one of the most important teachers of the small sword. Angelo's legacy also survived through his school of arms, which continued in his family for nearly 150 years.

Domenico Angelo was born in Leghorn, Italy, on 6 Febuary 1717. He was one of six brothers. Their father, Giovanni Tremamondo, was a leading merchant originally from Foggia in the Kingdom of Naples. Angelo's christened name was Angiolo Domenico Maria, but, as was common at that time, his legal name was buried under an array of titles and surnames. Historical records find him listed as Domenico Angelo Malevolti and Angelo Malevolti, among others. The use of the Malevolti family name stemmed from the family's claim to a Neapolitan marquisate. His mother, Catrina Angiola Malevolti, had continued to use her maiden name even after her marriage. Whatever title he may have personally preferred, as his life prospered in England he settled on the anglicised form of Domenico Angelo. His will was signed on 11 May 1797 as Domenico Angelo Tremamondo.

Angelo began his study of the sword at the academy in Leghorn, which was run by Andrea Gianfaldoni. Of course, the sword was only part of a gentleman's training, which also included the study of dancing and equestrianism. Angelo's studies at the academy did not go beyond the common schooling for a gentleman of the time. Travel was also an important part of a young man's life, and reports can be found of his visits to Florence, Naples, Rome, Turin and Venice throughout his twenties. By 1743 Domenico was well travelled but still had not decided upon a career. With the hope of encouraging his son to follow in his footsteps, Giovanni arranged for Domenico to move to Paris to study international trade.

In the middle of the eighteenth century Paris was the centre of the civilised world whence culture and fashion flowed throughout Europe. Domenico stayed there for at least a decade. However, his mind did not rest on merchant crafts and commerce for long, and he concentrated upon taking lessons in fencing, dancing and riding. It was the gentlemanly arts that finally provided Domenico with his chosen profession. He concentrated his energies upon equitation under one of the ablest teachers of the time, Monsieur de La Gueriniѐre, yet he did not neglect the other arts. Domenico continued to study all the gentlemanly disciplines, seeking out the best teachers in Paris. He was instructed in dance by the first dancer at the Opéra, Gaetan Balthasar Vestris, and fencing by Teillagory, reputedly one of the finest swordsmen of the time. Still living off his father's income, Domenico seemed very happy with his life in Paris.

He may very well have continued his Parisian existence for many more years had it not been for affairs of the heart. During a fencing demonstration, Angelo attracted the attention of Margaret Woffington, an actress visiting Paris. She presented him with a bouquet of roses from her corsage. The young gallant pinned the bouquet to his right breast and challenged all opponents that day to disturb any leaf of the arrangement. As in all truly romantic stories, the bouquet went the entire day undisturbed. A brief courtship period led to Domenico's quick decision to follow Peg (as Margaret was informally known) to England when she was called back to London. Shortly thereafter, Peg found employment in the company at the Aungier Street Theatre in Dublin, Ireland, under Thomas Sheridan. Of course, Angelo decided to join her, and in his time there he became good friends with many of Dublin's elite, including the Sheridan family. The couple stayed in Ireland until Peg's contract ended the following May and then returned to London. However, the passion between them had diminished, and upon returning to London Domenico fell in love with Elizabeth Johnson, the daughter of a deceased navy captain. Margaret seems to have even encouraged the courtship between them. In 1755, a little over a year after beginning their relationship, Domenico and Elizabeth were married at St George's Church in Hanover Square. The happy couple did not wait long to start their family. In 1756 they were blessed with the birth of a son, Henry, the first of their six children.

Angelo now had new responsibilities. Needing to take care of his family, he served as Master of the Horse for the Earl of Pembroke for several years after his marriage. In 1758, when the good earl was summoned by King George II for a demonstration of his prize horse, Monarch, Pembroke brought his riding-master along. At the end of the demonstration, His Majesty remarked that 'Mr Angelo is the most elegant rider in Europe'.[1] Such acclaim turned the heads of many nobles to Domenico's abilities. Soon after, the Dowager Princess of Wales offered Angelo the position of riding- and fencing-master for her sons, Edward the Duke of York and George the Prince of Wales (who would later become King George III). Domenico accepted the position and in 1759 moved his family to a suitable space in Leicester Fields which the dowager princess provided for him.

Although Domenico was hired to instruct in both fencing and riding, he had not previously taught the former, having chosen to concentrate on equitation. Shortly after beginning his work in Leicester Fields, Angelo received word that one of the leading Irish fencers, one Dr Keyes, wished to play a public bout with him. Angelo agreed to the meeting, and it was arranged to take place at the Thatched House Tavern in St James's St. On the day of the bout, Angelo arrived at the tavern still dressed in his riding clothes and found, to his surprise, that the room was filled with London's elite. He had expected a friendly bout with the Irish fencer, not a display for nobility and gentlemen.

Angelo quickly changed his state of mind and prepared for his fencing bout, even giving his excuses for his poor dress. Once both men were ready, they came to their guards. Keyes's style was much wilder and more aggressive than Angelo's calm and technical form. The spectators could easily observe that Angelo remained in full control of the bout. His calm manner only incited Dr Keyes to use beats with his blade and foot as well as to shout in an effort to throw his opponent off. However, nothing shook Angelo's control, and with a quick turn of the wrist the doctor's foil went flying as Angelo disarmed him. Recovering his foil, the doctor redoubled his attacks with even more vigour. Angelo easily parried these wild attacks and finally decided to end the bout. He touched the Irishman six times on the chest and finished off by disarming him again, this time using a different technique.

Needless to say, Angelo's victory was very popular with the crowd, and shortly after this display many powerful people pressed him to open a school of fencing. However, Angelo was quite happy instructing equitation. It took much persuasion from his old friend Pembroke, as well as his wife's friend the Duke of Queensberry, before Domenico finally consented to open a school of arms that also encompassed equestrian instruction. In order to facilitate this combined school, Angelo purchased the Carlisle House from Lord Delaval on 29 September 1761. The deed assigns the property to an Archibald Turner, who must have kept the property under his name because Angelo was still legally a foreigner; without receiving his Letters of Denization, Domenico could not own property. Once the

Carlisle House had been purchased, Angelo began work on transforming it into a school.

He and his family moved to their new residence and school in 1763. Throughout the end of the eighteenth century, a plethora of the English upper echelons would pass through the doors of Carlisle House for instruction in riding, fencing or both. The Angelo family grew to be well known and respected, and Carlisle House became renowned for top instruction in fencing and equitation. Meanwhile, George II had died in 1760, and now Angelo's most famous student, George the Prince of Wales, sat upon the throne as George III. It was George and his brother Edward who pressed Domenico to work on publishing a book.

The first edition of Angelo's *L'Ecole des armes* was published in 1763 by R. & J. Dodsley, the forty-seven illustrations provided by J. Gwyn Delin. The expenses of publishing at this time were often defrayed by a wealthy patron, but instead of just one benefactor Angelo's efforts were supported by 236 noblemen and gentlemen, among them some of the most prominent men of the time. The book was written in French and was bound in boards with a leather spine and a red label for the title. Several copies were printed for the king, princes and other nobility; these were large-format copies bound in green morocco and with the same red title label.

Angelo's decision to publish his book in French may be found confusing by some, considering his residence and that the distribution of the book was in England. It must be remembered that French was a language used by the cultured elite at this time, so gentlemen and courtiers were fluent in it. It is not surprising, though, that in less than two years a second edition was produced which had an English translation running in a column parallel to the French. The Dodsleys' imprint was replaced by S. Hooper in this 1765 edition, which proved so successful that another edition was printed in 1767.

Angelo's decision may also have had something to do with a project that had started in France in the previous decade. Denis Diderot and Jean le Rond d'Alembert had started compiling one of the largest encyclopedias in history. This twenty-eight volume *Encyclopédie* was published over the course of more than twenty years (1751–72). The seventeen volumes of text contained articles on everything from asparagus to the zodiac, and the final eleven volumes contained engravings that illustrated many of these articles. The work was a collaborative effort by 'men of letters and skilled workmen', and the gentlemanly art of fencing was certainly not neglected. When the duo started searching for an author for this section, there were many prominent masters to choose from – Danet,[2] Boëssière,[3] Gordine,[4] Menessiez,[5] O'Sullivan,[6] Massuet[7] and Weischner,[8] among many others. In the end, it was *L'Ecole des armes* that was chosen for inclusion. The editors introduced the text by saying, 'Had we known of any work more perfect in its kind, we should have made use of it.' This section of the *Encyclopédie*, entitled *Escrime*, wholly incorporated Angelo's treatise with only minor changes. Needless to say, the French masters were outraged with a native Italian, who was living and teaching in England, being chosen to represent the French style of fencing.

Angelo's treatise found its final incarnation in 1787 as *The School of Fencing*. Angelo's son Henry wrote a new preface to this edition. The book utilised new engravings, which were created in 1783 and are very close replicas of the original illustrations. It also excludes the original French text.

Much of Domenico's life can be recounted thanks to Henry Angelo's *Reminiscences*,[9] which contains much about his father's early life. Domenico's later years, however, are not nearly so well documented. By 1793 the seventy-six-year-old Domenico had begun to retire from the life he had built. He rented out Carlisle House and chose to live with his daughter, Sophia, in Eton, where he was the fencing-master for Eton College. He lived there happily until he passed away in 1802 at the age of eighty-six. He was buried in Windsor Parish Church. His lengthy obituary notice was published in the *Gentleman's Magazine*. Most likely composed by Henry, part of it reads:

> He retained his bodily powers so well that, at his very advanced age, he gave a lesson in fencing a few days before his death. He was a very respectable character, his manners were elegant and courtly, and he was well acquainted with life and familiarly known to most of the distinguished characters in Europe for the last half-century. He had long resided in this country, respected by persons of the highest rank, and particularly countenanced by the Royal Family.

Angelo's legacy is one of the greatest in the history of fencing. He started a lineage of fencing-teachers in England that would continue for over a century. *L'Ecole des armes* became a resource for swordsmen throughout all of Europe. Its concise and well-presented manner made it the most popular treatise on fencing for many years. It is quite rare for any fencing book to be reproduced four times in one decade as well as in two countries, and the longevity of

Angelo's treatise is matched by very few others. It remains an invaluable resource in the history of fencing and well worth our attention today.

---

[1] J. D. Aylward, *The House of Angelo*, London, 1953, p. 26.

[2] Danet (Syndic-garde de la compagnie des maîtres d'armes de Paris), *L'Art des armes*, Paris, 1767.

[3] La Boëssière, *Observations sur le traité de l'art des armes, pour server de défense à la vérité des principes enseignés par les maîtres d'armes de Paris, par M. maître d'armes des académies du Roi, au nom de sa compagnie*, Paris, 1766.

[4] Gérard Gordine (Capitaine et maître en fait d'armes), *Principes et quintessence des armes*, Liège, 1754.

[5] Menessiez, *Mémoire pour le sieur Menessiez, maître en fait d'armes, et maître des pages de M. le comte de Clermont. Contre la Communauté des maîtres en fait d'armes*, Paris, 1763.

[6] Daniel O'Sullivan (Maître en fait d'armes des académies du Roi), *L'Escrime pratique, ou principes de la science des armes*, Paris, 1765.

[7] P. Massuet, *La Science des personnes de cour, d'épée et de robe. Commensé par de Chevigny, continué par de Limiers, revue, corrigé et augmenté par P. Massuet*, Amsterdam, 1752.

[8] C. F. Weischner, *Exercices dans les salles d'armes*, Weimar, 1752.

[9] Henry Angelo, *Reminiscences of Henry Angelo*, London, 1830.


# ACKNOWLEDGEMENTS

I would like to thank the following people for their help and support:
Ramon Martinez, Suzanne Kirby, Ricki Ravitts, Ken Mondschein, Dave Storrs and Carol Crittenden.

# THE
# SCHOOL OF FENCING
WITH
## A GENERAL EXPLANATION
OF THE
## PRINCIPAL ATTITUDES AND POSITIONS
PECULIAR TO
# THE ART.

---

*By Mr. ANGELO.*

---

To their Royal Highneſſes the Duke of

# G L O U C E S T E R

AND PRINCE

# HENRY-FREDERIC.[†]

MOST SERENE PRINCES,

THE Honor already conferred on me, of teaching your Royal Highneſſes the Art of Fencing,[†] encourages me to preſent you with this detail on the ſubject; it is a ſlight tribute, for the many favours which you have condeſcended to beſtow

on

on me; might I presume still to form a wish, it should be, that your Royal Highnesses would acquit me of the idea of temerity, by permitting me publickly to testify the deep sense of gratitude, and the profound respect with which I remain of

YOUR ROYAL HIGHNESSES,

The most humble,

And most obedient Servant,

ANGELO.

# P R E F A C E.

---

WHEN the Goths had introduced the custom of single combat, the art of defence became a necessary study: it was confined to certain rules, and academies were instituted to train up youth in the practice of them.

THE moderns having adopted the small sword in preference to the ancient arms, it gave rise to a new species of defence, distinguished by the appellation of Fencing, which justly forms part of the education of persons of rank; giving them additional strength of body, proper confidence, grace, activity, and address; enabling them, likewise, to pursue other exercises with greater facility.

NOTWITHSTANDING this art has been carried, in practice, to so high a degree of perfection, few enquiries have been made into the theory of it; many French

and

and Italian maſters have communicated to the public their reflections on the ſubject, but they have not ſufficiently inveſtigated the moſt intereſting parts of it: this conſideration has induced me to compoſe and publiſh the following work.

I HAVE endeavoured to explain the principles and rules of the art in a ſuccinct and eaſy manner; I have given a circumſtantial detail of the different attitudes† of the body, and motions of the hands, arms and legs; and have, finally, added ſuch reflections and reſearches, that both the theorical and prctiacal parts will thereby elucidate each other.

I SHALL have attained to the accompliſhment of my wiſhes, if this work is ſo fortunate as to pleaſe a nation that I have been long devoted to, and which I ſhall always think myſelf happy in ſerving to the extent of my abilities.

TO THE

TO THE

# READER.

HAVING, under the auſpices of my father, and with his permiſſion, undertaken to give this edition of his Treatiſe on the Art of Fencing, I have endeavoured to render it of more general uſe, by reducing it both in ſize and price. As I follow the ſame profeſſion, in which my father has ſo highly diſtinguiſhed himſelf, I might be permitted to ſpeak to the merit of the principles laid down in this book; yet, as his ſon, my teſtimony might be called in queſtion. I ſhall, therefore, inſtead of my own, deliver the opinion of that learned body of men, the compilers of the French Encyclopedia†, whoſe judgment in matters of arts and ſciences cannot be ſuſpected of adulation or partiality.

UNDER

UNDER the article ESCRIME (Fencing) they ſpeak of the following ſheets, in theſe words:

"*This article is entirely taken out of a* TREATISE *on the* ART *of* FENCING, "*publiſhed in London by* MR. ANGELO; *we are indebted to him both for the* "DISCOURSES *and* PLATES. *Had we known of any work* MORE PERFECT IN "ITS KIND, *we ſhould have made uſe of it, &c.*"

SEVERAL French and Italian maſters have (as it has been obſerved before) ventured their thoughts on the art, but none of them ſufficiently expatiated on its material points: this conſideration has led me to publiſh this ſmall edition of a treatiſe ſo juſtly approved of, and ſo generally uſeful to the lovers of fencing.

FENCING ACADEMY, Opera Houſe,
Haymarket.

*H. ANGELO.*

THE

THE

# SCHOOL *of* FENCING.

## *THE METHOD OF MOUNTING A SWORD.*†

YOU muſt obſerve not to file or diminiſh the tongue† of the blade, for on that depends the ſtability and ſtrength of your ſword.

IF the tongue is too big for the mounting, you ſhould open the mounting; ſuch as the gripe, ſhell and pummel, and tighten the tongue, by putting in ſplinters of wood, ſo as to render it firm. The pummel and button muſt be of two pieces; the button ſhould be faſtened with a hollow ſcrew, four or five times on the tongue of the blade, which is to be run through the pummel, and rivetted according to the ſhape of the button, round or flat.

THIS is the beſt method of mounting a ſword, and which I recommend to all ſwordſmen. You will find this method very uſeful alſo for broad-ſwords,† or half-ſpadoons,† commonly called cut and thruſts.

You muſt obſerve that the gripe of the ſword be put on quite centrical to the heel of the fort[†] of the blade, which ſhould have a little bend above the fingers, when in hand, and let the whole mounting be turned a little inward, which will incline your point in carte.[†] This way of mounting your ſword will facilitate your diſengagements, and give you an eaſy manner of executing your thruſts.

### *HOW TO CHUSE A BLADE, AND IT'S PROPER LENGTH.*

I THOUGHT it neceſſary, before I ſet down any rules for the uſe of the ſword, to premiſe a few words, not only how to mount a ſword, but likewiſe upon the choice of a blade; for, with a bad ſword in hand, bad conſequences may enſue, be the perſon ever ſo courageous, and active. Some are for flat, others for hollow blades;[†] whatever pains were taken with the former, I ſeldom or ever found them light at the point; it is therefore difficult to render them light in hand; I would, nevertheleſs, recommend the uſe of them in battle, either horſe or foot; but in a ſingle combat, the hollow blade is preferable, becauſe of its lightneſs, and eaſe in the handling.

A PERSON

A PERSON ſhould proportion his ſword to his height and ſtrength, and the longeſt ſword ought not to exceed thirty-eight inches from pummel to point.

IT is an error to think that the long ſword hath the advantage; for if a determined adverſary artfully gets the feeble of your blade, and cloſes it well, by advancing, it would be a difficult matter for him who has the long ſword to diſengage his point, without drawing in the arm, which motion, if well timed, would give the other with the ſhort ſword an opportunity of taking advantage thereof.

YOU ſhould not fail obſerving, when you chuſe your blade, that there be no flaws in it; theſe flaws appear like black hollow ſpots, ſome long ways, others croſs the blade; † the firſt of theſe are frequently the cauſe of the blade's breaking.

THE temper of the blade is to be tried by bending it againſt any thing, and it is a bad ſign when the bending begins at the point; a good blade will generally form a half circle, to within a foot of the ſhell, and ſpring ſtraight again; if it ſhould remain in any degree bent, it is a ſign the temper of that blade is too ſoft: but though it is a fault, theſe blades ſeldom break. Thoſe which are ſtubborn in the bending are badly tempered, often break, and very eaſily.

## OF THE FORT AND FEEBLE OF A BLADE.

THE fort and feeble are equally on both edges of a blade. The Fort extends from the Shell to the Middle of the Blade, and the Feeble from thence to the point. You cannot attend too much to this diſtinction, ſince the executive part of the art, depends on a proper knowledge thereof.

## THE FIRST POSITION TO DRAW A SWORD.

## PLATE I.

YOU muſt ſtand ſtraight on your legs,† with your body ſideways; keep your head upright and eaſy, look your adverſary in the face, let your right arm hang down your right thigh, and your left arm bend towards your left hip; your left heel ſhould be near the point of your right foot, the point of your right foot in a line with your knee, and directed towards your adverſary; and, holding your ſword towards the hook of your ſcabbard,† you muſt

*The first Position to draw a Sword.*

*Publish'd as the Act directs Aug.^t 1783.*

2

3

*Position for the Guard in Carte.*

*Position for the Guard in Tierce.*

*Publishd as the Act directs Aug.t 1783*

muſt preſent yourſelf in order to draw. In this poſition, fixing your eyes on your adverſary, bend your right arm and raiſe it to the height of your ſhoulder, and carrying your hand to the gripe of your ſword, which hold tight and firm,[†] turning your nails toward the belt, draw your ſword, raiſing your hand in a line with your left ſhoulder, and make a half circle, with vivacity, over your head, preſenting the point in a line to your adverſary,[†] but no higher than his face, nor lower than the laſt rib, holding your arm ſtraight, without ſtiffneſs in the elbow, or the wriſt; in preſenting thus the point, you muſt raiſe the left arm in a ſemi-circle, to the height of your ear, and ſingle your left ſhoulder well, that the whole body may be in a profile; which inſtruction cannot be too cloſely attended to.

### *POSITION FOR THE GUARD IN CARTE.*

## PLATE II.

IN order to acquire this poſition, the left knee muſt be bent, and at two feet diſtance from the right; the left heel in a ſtraight line with the right heel, and the point of the foot perpendicular to the knee: you muſt obſerve that the bend on the left ſide ſhould not in the leaſt take off from that eaſe with which the body ought to be ſupported; and, to render yourſelf firm, bend the right knee a little, but not too much, for, if it is too much bent, the body might fall forward, and if not bent at all, neither the thigh, nor the leg, would be flexible,[†] and you would, conſequently, not have ſufficient elaſticity nor ſtrength to longe,[†] nor agility to advance or retire.

THE Guard in carte is the moſt advantageous, and the moſt elegant poſition in Fencing. There are in this art five different poſitions of the wriſt, offenſive and defenſive; which are, Prime, Seconde, Carte, Tierce, and Quinte. The two firſt to begin with are carte, and tierce; from which derive carte over the arm, low carte, and flanconade.

THERE are alſo in Fencing three openings, or entrances, viz. inſide, outſide, and low parts[†] of both theſe.

THE inſide comprehends the whole breaſt, from the left to the right ſhoulder.

THE outſide, all the thruſts made above the wriſt, on the outſide of the ſword.

THE low parts embrace all the thruſts made under the wriſt, from the arm-pit to the hip, from the inſide or the outſide.

The carte which is within, ought to be thruſt with your nails upward, and the inſide edge of the ſword a little more raiſed than the outſide one.[†]

A tierce ought to be thruſt on the outſide of the adverſary's ſword, with the nails downward, and the two edges of the ſword of equal height.

The prime ſhould be thruſt within both ſwords,[†] with the nails downward, and the edges of equal height.

The carte over the arm[†] ſhould be thruſt with the nails upward, and both edges of the ſword at an equal height.

The low carte[†] ſhould be thruſt below the wriſt, the edge being turned the ſame way as the carte within the ſword.

The ſeconde[†] ſhould be thruſt under the wriſt, with the nails downward, and the edges of the ſword of equal height.

The quinte[†] muſt be thruſt with the nails upward, directing your point to the outſide of your adverſary's wriſt, and elbow, and the edges of equal height.

The flanconade ſhould be thruſt from the inſide to the outſide of the adverſary's ſword, binding his blade, to convey your's under his elbow to the body, with your nails upward.

## *TO GET WITHIN, OR WITHOUT DISTANCE.*

IN order to come to our former poſition of the guard, it is very neceſſary to know what is meant by diſtance.

To get within diſtance, is called advancing on the adverſary, when he is at too great a diſtance from the point of your ſword; to be without diſtance, is to retire when your adverſary's point is too near.[†]

To get within diſtance of your adverſary without altering the regular poſition of your Guard, you muſt raiſe your right foot juſt above the ground, and carry it about a foot forward, in a ſtrait line with your left heel, bending your knee a little, and at the ſetting down of the foot, you muſt follow with the left leg in the ſame manner and diſtance, keeping your left leg well bent, to ſupport the body entirely on that ſide.

To get out of diſtance you muſt retire with the left foot, and follow regularly with the right foot, keeping always two feet diſtance (more or leſs according to your ſize) from one heel to another: you muſt be very cautious not to loſe your perpendicular poſition of body

and

and guard, elſe, by the diſorder which your adverſary could cauſe in your legs, the body would no more be firm, neither would the wriſt be able to execute with advantage when occaſion ſhould require it.

THERE is alſo a double advance,[†] by bringing up the left foot to the right toe, and the reverſe in the retreat, by carrying the right foot to the left heel; you may alſo jump back about two feet,[†] and, though this method is much uſed, yet I do not adviſe it, except you were on very level ground.

## *POSITION FOR THE GUARD IN TIERCE.*[†]

## PLATE III.

TO execute the tierce (as before mentioned) your nails muſt be downward, and engaging your adverſary's ſword, touch his blade; you muſt engage your point from inſide to outſide, by changing the poſition of your wriſt; ſo that when your wriſt is turned in carte, on the inſide of your adverſary's blade, you muſt, by a motion of the wriſt, drop your

your point close to his blade, turning your nails downwards; which is disengaging from carte to tierce.†

BEING in tierce, you must likewise drop your point, turning your wrist with your nails upward, and close your adversary's blade; which is disengaging from tierce to carte.

YOU must make frequent disengagements in this manner, in a firm position,† till your adversary retires, at which time you must disengage, and advancing, close his blade, with your point in a line to his body, always steady on your guard.†

WHEN you have thus disengaged, and advanced on your adversary, in these two positions, you must retire, and every time he disengages, you must turn your wrist on the side you are engaged; this will teach you to compleat your parades, in which the wrist is only to act: you must always oppose your adversary's sword sufficiently to cover the side he attacks, and you must nevertheless observe, that, when you cover one side, you do not uncover the other side or the lower parts.†

POSITION

*POSITION FOR THE INSIDE GUARD CALLED CARTE, AND THE INSIDE THRUST, CALLED THE THRUST IN CARTE.*

## PLATE IV.

TO execute this thruſt well, three motions of the wriſt are to be made at once; which are, to turn the wriſt and nails upward, raiſe the wriſt, and oppoſe;† and in theſe motions the arm ſhould be ſtraitened, and the wriſt raiſed above the head, and the point dropt in a line to the adverſary's breaſt; being thus ſituated, you muſt throw your wriſt forward, ſtepping immediately, or longe about two feet beyond your guard; the left heel, and knee, ſhould be in a perpendicular line, the point of the foot in a line to the knee, and the right heel in a line to the left;† the left foot ſhould be plumb to the ground,† and not move, heel or toe. And obſerve, that when the arm ſtretches forth in order to thruſt, the foot muſt follow at the ſame time; the body ſhould be very upright, the left leg ſtretched, and the left hand ſhould hang down in a line with the left thigh, about one foot diſtance, with the hand open, nails downward, and fingers cloſe.

THIS

*Position for the inside-Guarde call'd Carte & the inside Thrust call'd the Thrust in Carte.*

*Publish'd as the Act directs Aug.[st] 1783.*

THIS poſition of the right hand is to be obſerved in every thruſt made in carte,[†] the hand or wriſt ſhould go off firſt, and the point ſhould touch the body, before the foot is ſet to the ground; and, to perfect this thruſt, when the hand moves, the reſt of the body ought all to move with the ſame vivacity: and though it appears, when well executed, that the motions of all the parts are inſtantaneous, yet you will find the point has the priority.

OBSERVE well, that the body be firm, the head kept up, the left ſide from the hip well turned in, the ſhoulder eaſy, and the wriſt oppoſed to the ſword; that the pummel be directed in a line with your left temple, to prevent a counter thruſt from the adverſary's inſide, which will certainly happen without this oppoſition.

THE thruſt being made, the recovery to the guard muſt immediately follow, with the ſword in a ſtrait line[†] with the adverſary's body. You cannot practiſe this thruſt too much, it being the moſt eſſential and the moſt ſhining one that is made in fencing.

### *POSITION FOR THE OUTSIDE GUARD CALLED TIERCE, AND THE THRUST IN TIERCE.*

## PLATE V.

TO deliver this thruſt, your wriſt muſt be turned with the nails downward, and in the ſame height as in carte; the head muſt be covered by the oppoſition of the wriſt, though not in a line with the face; the inſide of the arm in a line with the right temple, the left arm to fall down about a foot from the thigh, the nails upward. Obſerve, that at all times when the right arm is turned with your nails down, that the left ſhould be the ſame,† and at the ſame diſtance from the thigh as in carte.

THERE are many fencers who, in delivering his thruſt, keep the wriſt in a line with the ſhoulder, and ſtoop with the head, to cover themſelves from a counter thruſt: in carte, likewiſe, leaning the head on the right ſhoulder. This not only hinders a ſight of the point, but renders it impoſſible to ſee ſo clearly as to prevent the adverſary's return, by a quick parry; for the head being in continual motion to ſeek ſhelter, and not knowing that the wriſt

*Position for the outside-Guard call'd Tierce, & the Thrust in Tierce.*

*Pub.d as the Act directs Aug.t 1783.*

*Position for the outside Guard call'd Tierce & the Thrust on the same side w.th the wrist revers'd in Carte call'd Carte over the Arm.*

*Publish'd as the Act directs Aug.t 1783*

wriſt is to cauſe the oppoſitions, they throw themſelves from the centre of gravity; and with a wavering body the delivery of the thruſt becomes ſtiff and aukward, and the recovery of their guard alſo. They are likewiſe liable, by bringing the body forward, to be expoſed to the adverſary's point.

*POSITION FOR THE OUTSIDE GUARD, CALLED TIERCE, AND THE THRUST ON THE SAME SIDE, WITH THE WRIST REVERSED IN CARTE, CALLED CARTE OVER THE ARM.*

## PLATE VI.

THIS thruſt muſt be delivered on the outſide of the adverſary's ſword, with the nails upwards as in carte, but in the tierce line, plunging the point to the adverſary's body; the wriſt ſhould be ſtrait, neither inclined to the outſide or the inſide, but raiſed, that the wriſt and pummel may come in a line with your right temple, the thumb and nails, and the flat of your blade in one line, and the other parts in the ſame poſition as in the thruſt in carte.

*POSITION*

*Position of the outside Guard call'd Tierce, & the Thrust from the outside under the wrist call'd Seconde.*

*Publish'd as the Act directs Aug.t 1783*

## *POSITION OF THE OUTSIDE GUARD CALLED TIERCE, AND THE THRUST FROM THE OUTSIDE UNDER THE WRIST, CALLED SECONDE.*

### PLATE VII.

THIS thruſt doth not differ from the tierce; but, becauſe it is delivered under the wriſt quite along the elbow, therefore the adverſary's ſword ſhould be engaged in tierce, dropping the point with the wriſt in tierce, directing it between the adverſary's arm-pit and his right breaſt; here the body ſhould be more bent forward, than what is mentioned in the former thruſts.

## *POSITION OF THE INSIDE GUARD CALLED CARTE, AND OF THE THRUST UNDER THE WRIST ON THE SAME SIDE, CALLED LOW CARTE.*

### PLATE VIII.

TO deliver this thruſt well, you muſt engage the adverſary's ſword in carte, dropping your point under his wriſt, in a line to his elbow, and in thruſting, not only fix your point

*Poſition of the inſide Guard call'd Carte & of the Thruſt under the wrist on the same ſide call'd Low-Carte.*

*Publiſh'd as the Act directs Aug.t 1783.*

point in his flank, but ſtrait traverſe the line about a foot outward,[†] without turning the foot to the right or to the left; forming an angle from the wriſt to the blade, the body as much bent as in the thruſt called Seconde, and the hand as much raiſed as in the carte thruſt: by this manner the oppoſition will be formed to cover the body and the face.

### *POSITION OF THE INSIDE CARTE, CALLED CARTE, AND THE THRUST GIVEN ON THE OUTSIDE FLANK, CALLED FLANCONADE.*

### PLATE IX.

TO execute this thruſt well, the ſword of the adverſary muſt be engaged in carte,[†] the point fixed in the flank of the adverſary, and, binding his blade, carried behind his wriſt, under his elbow. In this operation you muſt gain his feeble, and, without quitting his blade, plunge your point under his elbow to his flank, your wriſt turned nails upward, forming an angle from the wriſt to the point. In the execution of your thruſt, obſerve alſo, that the left hand ſhould drop under the right, and that too, form an angle, from the left elbow to the

*Position of the inside Carte, call'd Carte, & the Thrust given on the outside Flank call'd Flanconade.*

Publish'd as the Act directs Aug.t 1783

the wriſt, with your hand open, to prevent being hit on the parade of this thruſt, by the adverſary's turning his wriſt in tierce, and by thus reverſing his edge he would throw the point on you.

In the defenſive part of this parade I will explain this oppoſition.

## *THE SALUTE IN FENCING, GENERALLY MADE USE OF IN ALL ACADEMIES, AMONG GENTLEMEN, BEFORE THEY ASSAULT, OR FENCE LOOSE.*

THE ſalute in fencing is a civility due to the ſpectators, and reciprocally to the perſons who are to fence. It is cuſtomary to begin with it before they engage. A genteel deportment and a graceful air are abſolutely neceſſary to execute this.

### *FIRST POSITION OF THE SALUTE.*

### PLATE X.

YOU muſt ſtand on your guard in tierce, and, engaging the feeble of your adverſary's ſword, make three beats of the foot, called attacks, two of which are made with the heel, and the third, with the whole flat of the foot.

*1st Position of the Salute.*

*Publish'd as the Act directs Aug.t 1783*

Carry your left hand gracefully to your hat, without ſtirring the head, which is to face the adverſary; and, the hat being off, you muſt obſerve the following rules.

### *SECOND POSITION OF THE SALUTE.*

### PLATE XI.

YOU muſt paſs your right foot behind the left, at about a foot diſtance; keep your knees ſtrait, the body ſtrait, and the head very erect; at the ſame time ſtretch out your right arm, and turn your wriſt in carte, raiſing it to the height of your head, as much to the right as poſſible, holding the point a little low. When you paſs the right foot behind the left, you muſt drop and ſtretch your left arm, holding your hat with the hollow upward, about two feet from your thigh.

2.[d] Position of the Salute.

3.[d] Position of the Salute.

Publish'd as the Act directs Aug.t 1783

*THIRD POSITION OF THE SALUTE.*

## PLATE XII.

WHEN you have ſaluted to the right, obſerve well that the wriſt be carried to the left, bending the elbow, and keeping the point of your ſword in a line to the adverſary's right ſhoulder. All the other parts of the body ſhould be in the ſame poſition as before mentioned.

*FOURTH POSITION OF THE SALUTE.*

## PLATE XIII.

WHEN the ſalute is made to the left, the wriſt muſt gracefully be turned in tierce, holding the arm and the point of the ſword in a line to the adverſary, and at the ſame time come to your guard, by carrying the left leg about two feet diſtance from the right;† and bending the left arm, put on the hat, in an eaſy and genteel manner, and place the hand in the poſition of the guard.

13 14

4.th Poſition of the Salute. 5.th Poſition of the Salute.

Publiſh'd as the Act directs Aug. 1783

### *FIFTH POSITION OF THE SALUTE.*

### PLATE XIV.

BEING thus engaged, in the poſition of your tierce guard,[†] you muſt repeat the three attacks, or beats of the foot, and, ſtraitening your knees, paſs your left foot forward, point outward, the heel about two inches diſtant from the point of the right foot; and ſtraitening both arms, turn both hands in carte, the left arm about two feet from the left thigh, the right arm in a line with the right eye, and the point of your ſword in a line to your adverſary.

NOTE, Theſe laſt motions are to ſalute the adverſary.

AFTER this laſt attitude, you muſt come to your guard again, in what poſition of the wriſt you pleaſe, either to attack, or receive the adverſary.

IF you ſhould find yourſelf too near your adverſary, after having made your paſs forward with your left foot, you ſhould immediately carry your left foot back, and come to your guard, to ſhun an unexpected ſurpriſe, and by that receive the firſt thruſt; it being allowed

for

for either party to thruſt as ſoon as each is in his reſpective guard, as it is apparently probable that the adverſary is in a defenſive poſition.

THE copper plates of the figures (as explained in this book) have, in all their reſpective poſitions, the foils in their hands, to teach young fencers how to fix their points to their exact direction.

IN the art of fencing, much depends on a quickneſs of ſight, agility in the wriſt, a ſtaunchneſs in the parades, and keeping a ſolid firmneſs in the centrical motion of the body when a thruſt is made.

IN parrying, to have the body reſt entirely on the left hip and leg, to be flexible in the whole frame; not to abandon yourſelf, or flutter, but to be firm on your legs; alſo to underſtand your diſtance on every motion. But you cannot come to perfect all theſe, without great practice by leſſons, and by thruſting tierce and carte, of which I will give an explanation, and a juſt method.

## *METHOD AND EFFECTUAL MEANS TO RENDER A FENCER ACTIVE, AND FIRM ON HIS LEGS, AND TO SHEW HIM HOW TO RECOVER AFTER THE DELIVERY OF A THRUST, EITHER CARTE, OR TIERCE.*

YOU ought to practice not only to make your thrufts with great quicknefs and vivacity, but alfo to deliver them with an elaftic difengagement and difpofition;† and the motion of the body fhould appear like divers fprings throughout the whole frame.†

IT is very effential to recover from your thrufts in the fame lively manner, to enable you to parry in cafe of a return, or repofte.†

To this effect, as foon as the fcholar is able to thruft with firmnefs of body and legs, inftead of coming to his regular guard, he muft carry his right foot to the left, and alfo the left to the right;† and, in order that the fcholar execute this well, the fencing mafter is to give his affiftance for that purpofe, till he is able to execute the fame with eafe himfelf.

THE carte fhould be thruft without the affiftance of the plaftroon,† and, inftead of recovering to your natural guard, you fhould, with great agility and eafe, carry the point of your right

right foot to your left heel, keeping your body ereſt, the head alſo, and the knees ſtrait; you will find this to anſwer the fifth poſition of the ſalute in plate XIV.†

THE ſecond motion is the moſt difficult: you muſt, after having made your thruſt in tierce, inſtead of the uſual recovery to your guard, carry the point of your left foot before the right, your left heel to your right toe.† This ſhould be done with great eaſe and lightneſs, to bring you immediately on your legs; and it will enable you to come to a guard, by carrying the right foot forward, or the left foot back. You muſt obſerve, in either of theſe laſt motions, to move but one leg.

THE maſter, in order to aſſiſt his ſcholar when he is on the longe, ſhould keep up his right, with his left hand, till he brings him to a firm poſition of body and legs;† this will facilitate his recovery to a guard with eaſe and quickneſs, and will diſpoſe him to the motion of the paſſes which are to be made, and of which I ſhall hereafter give an explanation.

IT is very neceſſary, when the ſcholar takes a leſſon, that the maſter be attentive to withdraw his plaſtroon often when the ſcholar thruſts; for it will prove dangerous to uſe him to plant them always, by which he would find a reſt for his wriſt and foil;† for he would always abandon his head and body, and, inſtead of directing his point to a proper line of the

the adverſary's body, his wriſt, and, conſequently, his point, would be delivered, without rule, to the lower part of the body, and naturally fall lower ſtill; and the greater inconveniency would be, that he would not be able to recover his guard, nor parry, in caſe of a return, or repoſte: whereas, if the maſter often baulks his ſcholar, by withdrawing his plaſtroon at the time he thruſts and expects to find a ſupport or reſt for his foil, it will give him eaſe to deliver his thruſt, and to come to a defenſive poſition again, by making him attentive to keep his wriſt and body in a proper line, without dependance; and, it will give him a proper diſpoſition alſo to throw his point in a proper line and direction.

THE ſix thruſts which I have before mentioned, may not only be made from the blade directly to the body, which ought to be done at one equal meaſured time, ſtrait to the body, but alſo by a beat on the ſword, an appel of the foot, by a glizade, or ſliding on the blade, by a ſimple diſengage, or by a diſengage and an appel together.

THE beat on the ſword is done by engaging the blade either in tierce or in carte, or carte over the arm; you muſt leave the blade about four inches, and beat on it in a ſmart and lively manner, and thruſt firm, and ſtrait to the body.

THE appel, or attack, is made by raiſing the foot about two inches from the ground, ſetting it ſmartly down again, and thruſting directly at the body.

THE

THE ſliding, or glizade, on the blade, is done by firmly engaging the ſword, bending the elbow, and raiſing the point, to gain the adverſary's feeble; bringing your wriſt about a foot forward, to put by his point, by a preſs on his blade, and a thruſt firm and ſtrait to the body.

THE ſimple diſengage is done when you are engaged in tierce or carte, by quitting the blade to the reverſe ſide, without touching it, and thruſting ſtrait to the body.

THE diſengage with the beat of the foot is done at once, and the ſame time you diſengage you muſt join the blade of the adverſary, make your attack, and thruſt ſtrait forward to the body.

YOU muſt obſerve that theſe three different motions, viz. the diſengage, the attack, and the thruſt, ſhould be executed as quick as you may ſay—one, two.

### *OF THE SIMPLE PARADES.*

EACH thruſt hath its parry, and each parry its return. To be a good fencer, it will not be ſufficient to ſtand gracefully on your guard, nor even to thruſt with great ſwiftneſs and exactneſs:

exactneſs: the chief point is to be well ſkilled in the defenſive part, and to know how to parry all the thruſts that ſhall be attempted to be made at you.

When you are compleat in the defenſive part, you will ſoon be able to tire your adverſary, and often find an opening to plant a thruſt. You ſhould apply yourſelf to make your parades[†] cloſe, and firm in the line, by holding your ſword light, from the gripe to the point.[†]

The body ſhould be profile, or ſingled out[†] to the left ſide; and the wriſt and the elbow ſhould be the chief actors.

## *OF THE INSIDE PARADE CALLED CARTE, AND THE THRUST IN CARTE.*

## PLATE XV.

THIS parade of carte within the blade, is made by a dry beat on the adverſary's blade, with the fort of your blade, and your inſide edge. You muſt throw your body back in a ſtrait line with his, and let your oppoſition to his blade be about four inches wide to the left; your arm ſhould be a little contracted, and the moment you have parry'd, preſent your point firm in a line to his breaſt, to enable you to make a return, or repoſte, quickly.

*The Inside Parade calld Carte, & the Thrust in Carte.*

*Publishd as the Act directs Aug.t 1783*

## *OF THE OUTSIDE PARADE, CALLED TIERCE, AND THE TIERCE THRUST, CALLED THE OUTSIDE THRUST.*†

### PLATE XVI.

THIS outſide thruſt, called tierce, is parried by the inſide edge, and the turning of the wriſt to an outſide with a ſtretched arm, oppoſing the blade with the wriſt; without leaving the ſtrait line, you muſt lower your point towards the adverſary's body, to enable you to return the thruſt on the ſame ſide.

You muſt alſo parry this thruſt, by bending your arm, and oppoſing your wriſt, and by keeping your point to your adverſary's right ſhoulder. By this means you will be able to return the thruſt under his wriſt, called the thruſt in ſeconde.

*The outside Parade call'd Tierce & the Tierce Thruſt calld outſide Thruſt.*

*Publiſh'd as the Act directs Aug^t 1783*

## *OF THE OUTWARD THRUST, WITH THE NAILS UPWARD, COMMONLY CALLED THE FEATHER PARADE, AGAINST THE OUTWARD THRUST, NAILS UPWARD, CALLED THE CARTE OVER THE ARM.* †

## PLATE XVII.

IN order to parry this thruſt on the outſide of your blade, you muſt oppoſe with your outward edge, the wriſt as in carte, nails upward, and your wriſt in a line with your right ſhoulder, and, with a ſtrait arm, oppoſe the adverſary's blade with the heel, or fort of your ſword.

THIS thruſt may be parry'd alſo, by drawing in your arm, holding your wriſt a little outward, with your point raiſed; which being done, you muſt ſlide, and preſs from the feeble to the fort of his blade, by which you will not only put by his point, but have a great chance to fling his ſword out of his hand.

OF

*The outward Thrust, with the nails upwards commonly call'd the Feather Parade ag.^t the outw.^d Thrust, nails upward, call'd the Carte over the Arm.*

*Publish'd as the Act directs Aug.^t 1783.*

*The outside Parade for the Thrust under the wrist calld the Thrust in Seconde.*

Publish'd as the Act directs Aug.t 1783

## *OF THE OUTSIDE PARADE, FOR THE THRUST UNDER THE WRIST, CALLED THE THRUST IN SECONDE.*

### PLATE XVIII.

YOU muſt parry this thruſt with the inſide edge, and raiſe your wriſt, in ſeconde, to the height of your right ſhoulder, your point low, and well maintained from fort to feeble, directing your point between the arm-pit and right breaſt of the adverſary; and alſo keep a ſtrait arm, in order to throw off his point.

## *OF THE HALF CIRCLE PARADE, OR THE PARRY AGAINST THE INSIDE THRUST UNDER THE WRIST, CALLED THE LOW CARTE.*

### PLATE XIX.

THIS parry of the half circle ſhould be made within the ſword, by a ſmart beat on the feeble of the adverſary's blade with your inſide edge; your nails muſt be upward, your arm ſtrait, your wriſt raiſed to the height of your chin, and the point low, but well oppoſed from fort to feeble.

*OF*

*The half Circle Parade, or the parry against the inside Thrust under the wrist call'd the low-Carte.*

Publish'd as the Act directs Aug.t 1783.

## *OF THE PARADE AGAINST THE BINDING OF THE SWORD, FROM THE INSIDE, TO THRUST IN THE FLANK, CALLED FLANCONADE; BY REVERSING THE EDGE OF THE SWORD TO AN OUTSIDE, CALLED CAVEZ; AND AN EXPLANATION OF ANOTHER PARADE FOR THE SAME THRUST, BY BINDING THE SWORD.*

## PLATE XX.

THE reverſing the edge from an inſide to an outſide, called cavé, is a parade where you muſt, with great ſwiftneſs, turn your inſide edge to an outſide, at the very time the adverſary gains your feeble, by his binding, to direct his point in your flank, called flanconade, you muſt form an angle from your wriſt to your point, by which you will throw off the thruſt, and the point of your ſword will be in a line to the adverſary. You muſt keep a ſtrait arm, and maintain, with firmneſs, your blade, from fort to feeble.†

THE ſecond parade mentioned, called the binding of the blade, is made at the time the adverſary attempts to thruſt his flanconade. In order to this, you muſt yield your point, and ſuffer your feeble to be taken, ſo as to let your point paſs under his wriſt, without quitting his

*The Parade against the binding of the Sword from the inside to thrust in the Flank call'd Flanconade, by reversing the edge of the Sword call'd Cavez.*

*Publish'd as the Act directs Aug.t 1783.*

his blade in the leaſt, that your ſword may form a demi-circle; and, gathering his blade in carte,† you will find that the two ſwords, and wriſts, are in the ſame poſition as when the attack began, with only this difference, that the wriſts will be a little lower than in the ordinary guard.

### *OF THE PARADE CALLED PRIME, DERIVED FROM THE BROAD SWORD, AND CALLED THE ST. GEORGE GUARD, AGAINST THE OUTSIDE THRUST UNDER THE WRIST, CALLED SECONDE.*†

## PLATE XXI.

IN order to parry this outſide thruſt under the wriſt, called ſeconde, with this prime parade, you muſt, at the time your adverſary thruſts under the wriſt, paſs your point over his blade, and lower it to the waiſt, keeping your wriſt as high as your mouth, turning your nails downward, your elbow bent, your body kept back as much as poſſible, and give an abrupt cloſe beat on his blade with your outward edge; as you are then ſituated, you may, by

*The Parade call'd Prime deriv'd from the Broad-Sword & call'd the St George Guard against the outside thrust under the wrist call'd Seconde.*

Publish'd as the Act directs Aug.t 1783

by way of precaution, hang down your left hand, as before mentioned in the flanconade, and in the ſame manner,† or ſtep out of the line.

THIS oppoſition is made at the time you parry, and very cloſe, to avoid the adverſary's point, if you ſhould want to thruſt in a ſtrait line.

To ſtep out of the line, muſt be done at the time you parry the thruſt, by carrying your right foot, flat and plumb, about ſix inches out of the line to the right, the left foot alſo to be carried to the ſame line about a foot, which will throw you further from the centre.

IN my opinion, this laſt motion is preferable to the oppoſition of the left hand; and as it is practiſed in many academies, eſpecially in Italy, I have thought proper to give an explanation of it.

THE reaſon why I prefer this laſt to the firſt, is, becauſe the two points being low, and within the ſwords, it is better to ſtep out of the line; and by ſo doing, you will find the left ſide of the adverſary's body expoſed and open.

*OF*

### *OF THE PARADE CALLED QUINTE, THE POINT LOW, AND WRIST RAISED, AGAINST THE OUTSIDE THRUST UNDER THE WRIST, CALLED QUINTE THRUST.*

## PLATE XXII.

THE thruſt in quinte is made by making a feint on the half-circle parade, having your wriſt in carte. You muſt diſengage your point over the adverſary's blade, at the time he parries with the half circle, or prime parade, and thruſt directly at his flank.†

THIS thruſt is parried by holding your wriſt in high carte, with a low point, and by oppoſing from the forte of your outſide edge, to put by the adverſary's point; by a wriſt well maintained from fort to feeble, and a very ſtrait arm, having the body entirely ſupported by the left hip.

### *AN EXPLANATION OF THE VARIOUS THRUSTS THAT MAY BE PARRIED WITH THE FOLLOWING PARADES.*

ALL parades are made in general in the advance, the retreat, or by ſtanding your ground.

*The Parade call'd Quinte, the point low & wrist rais'd against the outside Thrust under the wrist call'd Quinte-Thrust.*

*Publish'd as the Act directs Aug.t 1783*

WITH the carte, by holding your wriſt low, you parry the low carte and the ſeconde; by raiſing your wriſt, you parry all the cuts over the point on the inſide of the ſword and the flanconade.

WITH the tierce you parry the carte over the arm; in raiſing your wriſt, you parry the cuts over the arm, carte over the arm or tierce.

WITH the feather parade, that is, with your outward edge when your wriſt is turned in carte on the tierce line, you parry the tierce thruſt; in raiſing your wriſt, you parry the cuts over the point on that ſide.

WITH the parade of ſeconde, you parry all the lower thruſts, both inſide and outſide; ſuch as low carte, ſeconde and flanconade.

WITH the half circle parade, you parry carte, tierce, carte over the arm and ſeconde.

WITH the prime parade, you parry carte, low carte, and ſeconde.

WITH the quinte parade, you parry ſeconde and flanconade.

## *OBSERVATIONS ON THE PARADES IN GENERAL.*

A GOOD parade is as neceſſary and uſeful when well executed, as it is dangerous and fatal if done without judgement, and made wide and rambling.

To parry well, will prevent your being hit; therefore you ſhould obſerve, when you are defending the place in which you are attacked, that you do not give an opening on the contrary ſide, which would give more eaſe to your adverſary to throw in a thruſt; for which reaſon you ſhould not flutter, or ſhew the leaſt concern, by any motion he may make, either with the body, his foot, or the point of his ſword.

THERE is not the leaſt doubt but you have a great advantage in forcing your adverſary to be on the defenſive, becauſe at this time it will be impoſſible for him to attack; and by this you will certainly find your account, by the openings he may through ignorance, or inadvertency give. And it is alſo very certain, that, by being able to baffle† his attacks, by a cloſe parade, your repoſtes, or returns, will be ſafe and quick, and according to the rules of fencing.

## *OF THE RETURN, OR REPOSTE, AFTER THE PARRY.*

EVERY parry hath its return; you will be reckoned a good fencer, when you parry with judgement, and return with a lively exactnefs.

THERE are in fencing two ways of returning a thruft; the one is, when the adverfary thrufts; and the other, when he is on his recovery to his guard.

THE firft of thefe is for thofe only who are well fkilled in this exercife, becaufe it requires a moft exact precifion, a quick fight, and a decifive parade; fince the adverfary ought to receive the thruft, before he has finifhed and executed his; which is termed, in fencing, a return from the wrift.

THE fecond, which is on the adverfary's recovery to his guard, is done by hitting him before he has fet his right foot to the ground again, thrufting out with great celerity and refolution, in order to execute well.

THE return, called the return of the wrift, fhould hit the adverfary at the very time he longes out to thruft. This method muft be executed with the greateft quicknefs poffible.

IN

In order to ſucceed, you muſt be firm on your legs, and, after having parried with the fort of your ſword, in a dry and abrupt manner, you muſt ſtraiten your arm, and bring your body a little forward on the right leg, remarking, attentively, that your wriſt direct your point to the adverſary's body; as you may obſerve in the twenty-fourth, twenty-fifth, and twenty-ſixth plates.

## *OF THE RETURN IN CARTE, AFTER THE CARTE PARRY.*

## PLATE XXIII.

AT the very time the adverſary delivers his carte thruſt, you muſt parry with the heel of your blade, and inſtantly return the thruſt within the ſword, and return to your guard as quick, according to the rules already explained.

If the adverſary ſhould in the leaſt raiſe his wriſt on his recovery, you may return a low carte, and recover with your wriſt in tierce, or demi-circle.†

*The return in Carte, after the Carte-Parry*

*Publishd as the Act directs Aug.t 1783*

### *OF THE RETURN IN TIERCE, AFTER THE TIERCE THRUST.*

## PLATE XXIV.

AT the time that you parry the tierce with a ſtrait arm, and your point a little lowered to the adverſary's body, you muſt return the ſame thruſt, only your wriſt a little inclined to the outſide. Take great care that the hand moves firſt, and oppoſe his blade well, from feeble to fort; recover to your guard in prime, or demi-circle parade.

You may alſo, after your tierce parry, return the thruſt in ſeconde, and recover in ſeconde, demi-circle, or in tierce.

### *OF THE RETURN IN SECONDE, AFTER HAVING PARRIED CARTE OVER THE ARM.*

## PLATE XXV.

AFTER your parry carte over the arm, you may return the ſame thruſt, by holding your wriſt nails upward, thruſting and oppoſing with your outward edge, and plunging your point to his body, with your recovery to a half-circle parade.

If

*The return in Tierce after the Tierce thrust.*

Publish'd as the Act directs Aug.t 1783

If you ſhould return a ſeconde thruſt, you ſhould, the moment you have parried carte over the arm, drop your point along the outſide of his wriſt and elbow.

This return is eaſier to be made than the aforementioned; your recovery is a half-circle, ſeconde or tierce.

## *OF THE RETURN IN QUINTE, AFTER THE THRUST IN SECONDE.*

## PLATE XXVI.

HAVING parried the thruſt in ſeconde with the quinte parade, you muſt return the thruſt without altering your wriſt. If you parry with the parade in ſeconde, you muſt return in ſeconde, recover ſwiftly with your wriſt in ſeconde, and bind the blade in carte, after the recovery, without leaving it.

*The return in Quinte after the Thrust in Seconde.*

Publish'd as the Act directs Aug.t 1783.

*The return in Seconde after having Parried Carte over the Arm.*

*Publish'd as the Act directs Aug.^t 1783*

## *OF THE RETURN ON THE FLANCONADE THRUST, BY REVERSING THE EDGE, TO THOSE WHO HAVE NOT THE PRECAUTION TO OPPOSE WITH THE LEFT HAND.*

### PLATE XXVII.

AT the time the adverſary thruſts the flanconade, you muſt parry, by raiſing and turning your wriſt in tierce, without leaving his blade, forming an angle from wriſt to point, ſteadily directed to his body.† In this return there is no occaſion to thruſt out the right leg; you muſt only bring your body forward, and ſtretch out your left leg.

THE angle which is formed in turning the wriſt is quite ſufficient to keep off and return the thruſt; this done, you muſt recover your guard in prime, or half circle.

IF you parry the flanconade, by binding the ſword, as I have before mentioned, you may return the thruſt ſtrait in carte; and if, on the adverſary's recovery, he ſhould in the leaſt drop his point, you may return a flanconade; if, on the contrary, he ſhould raiſe his wriſt or his point, you muſt return a low carte, and recover by a circle,† joining his blade.

*OF*

*The return on the Flanconade thrust by reversing the Edge, to those who have not the precaution to oppose with the left hand.*

Publish'd as the Act directs Aug.^t 1783

*The return from the Prime Parade to the Seconde & Low Carte Thrust.*

*Publish'd as the Act directs Aug.t 1783*

## *OF THE RETURN FROM THE PRIME PARADE, TO THE SECONDE AND LOW CARTE THRUSTS.*

## PLATE XXVIII.

AS ſoon as you have parried the ſeconde, or the low carte, thruſt with the prime parade; ſtepping out of the line with the right foot (as explained in the parades thereof) you muſt return the thruſt in prime, holding your wriſt in a line with your left ſhoulder; by this means you will form an oppoſition which will cover your body and face; after which you muſt recover in the ſame poſition of prime, or in the half circle.

## *OF THE PARADE BY A COUNTER DISENGAGE.*

THIS parade is made at the time the adverſary delivers his thruſt, by diſengaging carte or tierce, or carte over the arm: therefore, to execute this parade well, you muſt, the moment he diſengages to thruſt, diſengage alſo, very cloſely to his blade, and, having formed your parade, which ought to be done with the greateſt preciſion and quickneſs that is poſſible, ſupport your blade from fort to feeble.†

*OBSERVATION*

## OBSERVATION ON THIS PARADE.

AS a quick wriſt and a light point might eaſily deceive this laſt counter diſengage; that is to ſay, if the adverſary ſhould continue his diſengages often, and quicker than you can follow him, in ſuch a caſe, to ſtop his career, you muſt have recourſe to the circle parade, which will certainly ſtop the progreſs of his point.

## OF THE HALF CIRCLE PARADE.

THIS parade, which is the chief defenſive parade of the ſword, parries not only all the thruſts, but alſo obſtructs all the feints that can be made; and, to execute it well, you ſhould ſtraiten your arm, keep your wriſt in a line with your ſhoulder, your nails upward, and, by a cloſe and quick motion of the wriſt, the point ſhould form a circle from the right to the left, large enough to be under cover from the head to the knee; in this manner, by doubling your circle till you have found the adverſary's blade, your parade will be formed.

AND now, in order to ſtop this circle parade, notwithſtanding its being redoubled with great

great vivacity, you may ſtop his blade ſhort, by keeping your wriſt the height of your ſhoulder, and lowering your point, as in the quinte parry; and, recovering, bind and gather his blade in carte.

You ſhould exerciſe and practiſe theſe circle parades, from the counter diſengages to the circle, and from the circle to the counter diſengage. You may practice this leſſon yourſelf, either with ſword or foil: this will ſtrengthen and ſupple your wriſt, and will inſenſibly procure great eaſe and readineſs to defend yourſelf upon all occaſions.

## *METHOD OF THRUSTING AND PARRYING TIERCE AND CARTE, CALLED THRUSTING AT THE WALL.*†

IT is called thruſting at the wall, becauſe the perſon who parries is not to move his body nor his legs from where he ſtands; it is only his wriſt which is to perform his parades.

IN order to execute this leſſon well, I will explain the poſition in which he that parries is to ſtand.

To parry at the wall, you muſt place yourſelf ſo that the left foot may not be able to ſtir further; you muſt ſingle your ſhoulder, hold your head erect, pull off your hat, and open with

with your right arm, ſo as to carry your point to the right, that your adverſary may take a proper meaſure, or diſtance, for his longe;[†] after which you put on your hat, and carry your left hand back to the ſmall of your back, and giving an opening, either in tierce or carte, you muſt be ready to parry (in this attitude) with quickneſs, all the adverſary's thruſts.

To thruſt againſt the wall, you muſt place yourſelf ſtraight on your legs, as in the firſt poſition, or plate; and, coming to your guard, pull off your hat with a pleaſant countenance, and deliver a gentle thruſt in carte toward the adverſary, gently touching your button[†] to his breaſt, to take your diſtance; you muſt then recover to your guard, and put your hat on, making, by way of ſalute, the two motions of the wriſt, in tierce and carte, being the uſual way.

Thus, thruſting againſt the wall accuſtoms the ſcholar to thruſt with quickneſs, and to parry ſo likewiſe; it gives an exactneſs and eaſe, as well as a knowledge of diſtance, and is more uſeful, as in fencing with various perſons you will meet with people of different make and ſize.

*METHOD*

### *METHOD HOW TO THRUST AT THE WALL WITH SWIFTNESS.*

THERE are three different ways of thrufting at the wall. The firft is, by engaging the fort of the adverfary's blade, and holding a loofe point, you muft difengage lightly, and thruft ftrait at his body, feeking his blade.†

THE fecond is, by engaging from the point to his blade,† with a bent arm, difengage, and thruft ftrait to the body; which is called thrufting from point to point.

THE third is, by delivering ftraight thrufts to the infide and to the outfide of the adverfary's body; in this latter you need neither engage the blade, nor difengage.†

THE Italian mafters make much ufe of this laft, as it gives great fwiftnefs, and ufes the wrift to execute firft, and alfo loofes the fhoulder completely.

### *RULES TO BE OBSERVED IN THRUSTING AT THE WALL.*

WHEN you have taken your dimenfions, or diftance, as before mentioned, you ought in no way to ftir your left foot, nor the body, make no motions or feints whatfoever, but thruft according to rule, either infide or outfide, by difengagements, or by thrufting

ing to the blade ſtraight forward; and if you make any feints it muſt be with the mutual conſent of each other.

## *EXPLANATION OF WHAT IS CALLED FEINTS.*

A FEINT is, to ſhew the appearance of a thruſt on one ſide, and execute it on the other. In this you ſhould lead the adverſary's wriſt ſo much aſtray as to obtain an opening ſufficient to throw in the thruſt you have premeditated. You muſt be very cautious not to uncover yourſelf in making a feint; for, inſtead of ſucceeding in your project, you will give the adverſary an opportunity of a wide opening, and induce him to throw in a ſtraight thruſt: therefore it is abſolutely neceſſary, in making a diſengage in order to feint, to oppoſe the fort of your blade, and, with great ſubtilty, make the motion of your point near enough to his fort, that you may the more eaſily hit him.

ALL feints may be executed ſtanding ſtill, or in moving; you may make them after an appel, or attack of the foot, after a beat on the blade, or at the time the adverſary forces your blade; or at his diſengaging.

You

You muſt obſerve that, when you feint, your wriſt is the height of your ſhoulder, your elbow a little bent, that your wriſt may be more flexible, and your point lighter.

If you are engaged in carte, diſengage carte over the arm, near the adverſary's fort, bring your point back in its carte poſition, thruſt ſtraight forward, and recover to carte, or in the demi-circle.

If engaged in carte over the arm, you muſt diſengage in carte, and thruſt carte over the arm, and recover in tierce, or in half circle; and if the thruſt is parried, you muſt treble[†] the feint, and thruſt within the ſword.

If engaged in tierce, you muſt mark your feint below the wriſt in ſeconde, thruſt in tierce, and recover to your ſword in ſeconde, or a half-circle.

If engaged in tierce, you may alſo feint in carte, keeping your point in the adverſary's face, thruſt a low carte, and recover to the ſword by a circle.

To feint on any diſengagement of the adverſary, you muſt force or lean on his blade a little, to oblige him to ſlip, or diſengage; and at that very time you muſt, with quickneſs and preciſion, feint on his motion, and thruſt ſtrait at him.[†]

DEFENCE

### *DEFENCE OR PARADE AGAINST ALL FEINTS.*

THE ſureſt defence againſt feints, is to get at the adverſary's blade by a counter diſengage, or by a circle; for if you offer to ſeek the blade by a ſimple parade, it will be impoſſible to get at it, ſince he can redouble his feints at pleaſure; wherefore, by the parades, you immediately ſtop the adverſary's point, and you oblige him to change his intention and idea.†

### *OBSERVATIONS ON THE FEINTS, AND AT WHAT TIME THEY ARE GOOD AND BAD.*

THERE are fencers who, when they feint, make large motions of their body, or their points, or violent attacks of the foot, in order to precipitate their adverſary's defence, thinking to take advantage of the large openings he may on ſuch occaſions give; all theſe attempts, which are falſe,† cannot ſucceed againſt any but thoſe who are timid, and not ſtaunch in their guards; but, againſt a man who is ſkilful and cool, and who keeps his point cloſe in in the line to his adverſary, and who ſeeks his blade with the wriſt cloſely,[†] according to the rules of fencing, whatever motions may be made by feints will prove ineffectual.

THERE are others who make feints by advancing their ſword, and when the parry is attempted, drawing and changing the point from its ſituation, thruſt out :[†] theſe three motions are contradictions to each other, and are ſo ſlow, that, if the adverſary was to thruſt at the time they draw in, they would infallibly be hit before they had finiſhed their feint and intent.

YOU ſhould, as much as poſſible, make all feints in proper diſtance, that you may be enabled to accompliſh your thruſt with ſwiftneſs.

YOU may alſo make the feints out of diſtance; but you muſt continue them at the time you advance to get into diſtance, and change your idea, if the adverſary ſhould come to join your blade.

YOU ought to cover yourſelf well in making theſe motions, for the adverſary might thruſt ſtraight forward at the time you advance, or ſtop your feint by any other motion.

YOU are not always to hope or expect your feints ſhould be anſwered, for by this you may eaſily be baulked; but, knowing the adverſary has power to attack, and keeping yourſelf on your guard, you will be more ready to defend yourſelf. You will alſo be the more ſure that your feint will beſt ſucceed while the adverſary's wriſt is in motion; at that time ſeizing the opportunity to make your feint, he will become apt to fly to the defenſive with more irregularity; and not only will give openings, but you are ſure he cannot attack; and, conſequently you will be apt to hit him with more eaſe, and without riſque.

### *OF THE MOTIONS MADE ON THE BLADE STANDING STILL, CALLED GLIZADES,† AND THE GLIZADE FROM CARTE OVER THE ARM, TO THRUST CARTE.*

IF you are engaged in carte, and are in diſtance, you muſt have a flexible arm, your body ſingled, and entirely on the left hip; in this poſition you muſt make a beat† on the adverſary's blade, with an intent to ſtir his wriſt; if he ſhould come to the ſword, you muſt diſengage lightly carte over the arm,† with your wriſt high, and your point in a line to his face; and, the moment he cloſes the blade, diſengage in carte, and thruſt directly ſtraight. If, after this, he ſhould not return, but only force your blade, you may reiterate a ſecond thruſt, by turning your wriſt in tierce, on the blade, without leaving it, and recover to his ſword in carte.

*OF*

### *OF THE GLIZADE FROM CARTE, TO THRUST CARTE OVER THE ARM.*

IF you are engaged in tierce, or carte over the arm, you muſt diſengage in carte, keeping your point forward in a line to the adverſary's body; at the ſame time make an attack with the foot, and when you find he forces your blade, ſlip, or diſengage quickly in carte over the arm, and thruſt out,† and redouble the thruſt in ſeconde, recovering ſwiftly to the ſword in tierce, or in demi-circle.

### *OF THE GLIZADE FROM TIERCE, TO THRUST THE SECONDE.*

IF you are engaged on the tierce ſide,† after beating on the blade, and an appel, you muſt diſengage to tierce, with your point advanced to the enemy's face, and the moment he feels the blade, parry,† either by tierce, or the wriſt turned in carte over the arm, you muſt drop your point under his wriſt, and thruſt ſeconde,† recovering your ſword in tierce, or the half circle.

## *OF THE GLIZADE FROM CARTE, TO THRUST LOW CARTE.*

IF you are engaged on the outſide of the ſword, you muſt endeavour to move the adverſary's wriſt from the line, by a beat of your fort on his feeble; and diſengaging your point in carte, keep it in a line to his face, ſtretching your arm, and making an attack of the foot, ſlip your point under his elbow, and thruſt the low carte; recover immediately your ſword in tierce, and redouble† a ſecond thruſt; after which, recover the ſword by a circle.

ALL theſe glizades are made without advance or retreat; may alſo be made in the advance, if you ſhould find your adverſary retreat on your beats, or attacks of the foot.

IF, after theſe laſt motions, the adverſary ſhould want to ſhun the point by his retreat; in that caſe you ſhould diſengage, and quickly advance, and put in execution the before mentioned operations; but always obſerve to keep your ſword well before you, and your body backward, that you may not be ſurprized at any time.

OF

## *THE BINDING AND CROSSING[†] THE BLADE.*

VERY few maſters teach the croſſing of the blade; this operation is the more neceſſary, being not only uſeful to put the adverſary's blade by from the wriſt to the knee, but will often throw his ſword out of his hand.

If he ſhould preſent himſelf in diſtance, having his arm ſtraight and his point alſo, you ſhould incline your body entirely on the left ſide,[†] and engage his ſword in carte, turning your nails a little more upward than in the ordinary guard, and engage your fort about a foot from the feeble, directing your point to the left; in this poſition you muſt ſwiftly turn your wriſt in tierce, holding your ſword firm, and paſſing your point over his arm, without quitting his blade, ſtop your ſword from fort to point, holding your hand the height of your ſhoulder, and the point as low as his waiſt; this croſſing being made, thruſt out at full length in tierce, and recover your ſword by a circle.

### *ANOTHER WAY OR MANNER TO CROSS THE BLADE IN CARTE.*

IF the adverſary ſhould make a carte thruſt, you ſhould parry it with the half circle parade, keeping your body as before, well on the left hip, and as ſoon as you have parried, turn, with agility and firmneſs, your wriſt in tierce, inclining your point on the left, and finiſh by preſſing his feeble with the fort of your tierce edge.

IF theſe croſſings of the ſword are made with ſwiftneſs and preciſion, it is almoſt impoſſible not to diſarm your adverſary, or at leaſt not to put by his blade, ſo as to enable you to throw in a thruſt in tierce, as I have before mentioned.

### *THE MANNER TO SHUN THESE CROSSINGS OF THE SWORD.*

AS every thruſt has its defence, it is neceſſary I ſhould explain them in the cleareſt manner; there needs only a due attention, a quick eye, and judgment, to ſhun all thruſts.

YOU muſt give way with your point by a diſengage to a ſtrait line, at the time the adverſary wants to croſs your blade; by this means, as he will find no blade, you will evade it.

It

It might happen, even that when he finds no blade, and consequently no resistance to his blade, he may fling his own sword away, instead of his adversary's, if he should not maintain his point to the waist, and hold it very fast.

### *OF THE BEAT ON THE TIERCE THRUST TO FLING DOWN THE SWORD OF THE ADVERSARY*

IF the adversary should make a full thrust in tierce, you must disengage nimbly in carte, have your body well on the left hip, and draw in your arm a little more than in the ordinary guard, with your point high: this done, before he recovers to his guard you must make a smart beat with your fort on his feeble; which will open his fingers and throw down his sword.

### *ANOTHER WAY TO BEAT THE SWORD OUT OF THE ADVERSARY'S HAND.*

IF the adversary thrusts in tierce, you must parry with the prime parade; when parried, raise your point lightly to the left, and with the carte, or inside edge of the fort of the blade, beat smartly and strong on the feeble of his blade.

*OF*

### *OF THE BEAT ON THE SWORD IN CARTE OVER THE ARM.*

IF the adverſary preſents himſelf, having his wriſt turned in carte, with a ſtraight arm, and his point in the ſame line, you ſhould make a cloſe feint toward his fort; if that ſhould not move his point, you muſt diſengage in carte over the arm, with your arm contracted, and your point high, and with the fort of your blade beat ſmartly on his feeble; ſtraighten your arm immediately, and thruſt firm and ſtraight carte over the arm. If his ſword does not drop by this beat, you will at leaſt find opening enough to throw in your thruſt; this done, recover quickly to your guard by the circle parade.

To ſhun this beat you muſt, the moment you have made your thruſt, recover to the adverſary's blade by the circle parade; for this beat on the blade is only made uſe of againſt thoſe who, when they have thruſt either in tierce or carte, within or without the blade, do not recover immediately to their guard; or when they are in guard, keep their arm ſtraight, and the point of their ſword in a direct line with their arm.

If the adverſary ſhould attempt to beat when you are firm on your legs, and in guard, you muſt ſlip his beat, and with ſubtilty hinder him from touching your blade, and at that very moment go on with your thruſt ſtraight forward.

*OF*

## *OF THE PLAIN CUT OVER THE POINT FROM CARTE, IN TIERCE.*

## PLATE XXIX.

AFTER a thruſt made in carte, if your adverſary parries with the feeble of his ſword, you muſt, on your recovery, draw in your arm, keep a ſtraight blade from fort to point, and paſs it over his point; your wriſt being turned in tierce, you muſt raiſe it inſtantly, and plunging your point to his body, you muſt make a full thruſt in tierce, and recover to the ſword in the line.

THIS cut over the point is alſo made after having parried a carte thruſt, if the adverſary is fearful of a return on that ſide, and therefore forces your blade in his recovery, you ſhould, in ſuch caſe, execute the cut over, ſmartly to an outſide, either in tierce or carte over the arm.

YOU may alſo execute the ſame cut over the point, in carte over the arm, after you have made a carte thruſt, if the adverſary ſhould force your blade ſtanding on his guard.

OBSERVE well, that all theſe cuts over the point are not made uſe of but where the adverſary

*The plain Cut over the Point from Carte in Tierce.*

*Publish'd as the Act directs Aug.t 1783.*

ſary parries from the half ſword to the point, called the feeble, or when he forces your blade out of the line in the ſame manner.

### *OF THE PLAIN CUT OVER THE POINT, FROM TIERCE TO CARTE.*

IF you are engaged in tierce, you muſt make an attack with the foot, and execute a half thruſt to the ſword; and at the time the adverſary comes to this parade in tierce, you muſt draw in your arm, keeping a ſtraight point, and cut over his point in carte, your wriſt turned in carte, dropping your point a little, and make a full thruſt in carte; the thruſt made, recover ſtraight in a line to the adverſary's ſword, with an oppoſed wriſt, and your point to his body.

### *CUT OVER THE POINT FROM TIERCE TO CARTE, TO THRUST CARTE OVER THE ARM.*

IF you are engaged on the outſide of the ſword, turn your wriſt as in carte over the arm, attack ſmartly with the appel of the foot, and at the adverſary's coming to a parry, cut over the point to the inſide, with your wriſt high in carte, and your point ſtraight to his body,

body, as if you intended to thruſt there; and as ſoon as he comes cloſe to your blade, diſengage lightly, and thruſt a full carte over the arm, by ſending your wriſt firſt, with a ſufficient oppoſition, as before mentioned in the rules of fencing: the thruſt made, recover to a guard with your ſword before you, and the point to the adverſary's body.

### *ANOTHER CUT OVER THE POINT, FROM CARTE TO CARTE OVER THE ARM, TO THRUST CARTE.*

IF you are engaged in carte, make a half thruſt to the ſword, to oblige the adverſary to come to the parade, and at that inſtant cut over, and ſeem to thruſt carte over the arm, and without delay diſengage your point under his wriſt, and thruſt carte, maintaining and oppoſing your wriſt well; after which, recover to the ſword in carte, firmly on your left hip

*ANOTHER*

### *ANOTHER CUT OVER THE POINT, FROM CARTE TO TIERCE, IN ORDER TO THRUST SECONDE.*

IF you are engaged in carte, make an appel of the foot, cut over the point from carte to tierce, turning your wriſt to tierce; and your point being paſſed over to the outſide of the ſword, without heſitation drop it under the elbow of the adverſary, and execute a full thruſt in ſeconde; the thruſt made, recover inſtantly to a tierce, or half circle parade.

YOU may double the cut over, from carte to carte over the arm, and in lieu of delivering the thruſt on the firſt, cut over to the outſide: you may cut over again, and thruſt carte.

IT may alſo be made when engaged on the outſide, by cutting over to the inſide, and redoubling the cut to the outſide, either carte over the arm, or tierce.

I MAKE very little uſe of theſe double cuts, becauſe the plain cuts are preferable by their ſwiftneſs, and conſequently more difficult to parry; nevertheleſs, I think it neceſſary that all ſwordſmen ſhould know there are ſuch thruſts in fencing.

OF

## *OF THE DEFINITION OF THE WRIST, AFTER THE THRUST MADE IN CARTE.*†

THIS definition of the wriſt is not made uſe of but to thoſe who do not return from the thruſts made at them, either to the inſide or outſide; to execute this, you muſt engage the adverſary's ſword in tierce, make an appel of the foot, and as ſoon as he cloſes your blade, diſengage lightly near his fort, and thruſt in carte; the moment your thruſt is made, ſeem to recover to a guard, by bringing up your right foot about a foot back, keeping a ſtraight, and yet a flexible arm, and at the very moment he lifts up his right foot to cloſe in, if even he ſhould force your blade, you muſt take the opportunity, and turning your wriſt ſwiftly in tierce, thruſt in this manner to his blade, forcibly and well maintaining your wriſt; this is called cavé; the thruſt finiſhed, you muſt recover to the ſword in a ſtraight line.

*OF*

### *OF THE DEFINITION OF THE WRIST, IN CARTE OVER THE ARM, AFTER THE THRUST IN TIERCE.*

IF you are engaged within the ſword, you muſt make an attack of the foot, and a beat on the blade. If the adverſary comes to the blade, diſengage nimbly to an outſide, with your wriſt turned in tierce, and deliver your thruſt in tierce; the thruſt made, immediately recover to your guard about a foot, and the moment he lifts up his foot to advance, you muſt ſeize that moment, and turning your wriſt on his blade, nails upward, thruſt carte over the arm; the thruſt made, you muſt recover to the ſword in tierce, or in half circle.

### *OF THE PASS† ON THE SWORD, IN CARTE OVER THE ARM.*

## PLATE XXX.

IF you find the adverſary engaged in tierce, with a bent elbow, and a high point, you muſt recall him by an attack of the foot, and make a half thruſt to his ſword; if, after this he ſhould not anſwer your motion by coming to your ſword, then without altering the poſi-tion

*The Paſs on the Sword in Carte over the Arm.*

Publiſh'd as the Act directs Aug.t 1783

tion of your right foot, bring up your left foot, and paſs it before the right, about two feet forward, the point of the foot a little outward, and raiſe the heel of your right foot a little, ſtretch out both knees, and let the weight of the body be on the left leg;† and obſerve, that when you paſs your left leg forward, you turn your wriſt in carte over the arm, and with a ſtraight arm thruſt firm to the body; the moment you have made your thruſt, you muſt recover your guard in tierce, carrying, at the ſame time, your left leg, your body and arm to the firſt poſition, and ſeek his ſword with a circle parade.

### *OF THE PASS IN TIERCE ON THE OUTSIDE OF THE SWORD.*

IF the adverſary is engaged in carte, you make a ſtrong beat with your fort on his blade, and if, after you have put by his blade, he comes to a forced parry, diſengage ſubtilly to his outſide, with your wriſt raiſed in tierce, and the point plunged to his body; and at the very time of your diſengaging, bring up your left foot before the right about two feet, keeping the knees ſtraight, and the body reſting on the left leg; the thruſt made, recover to the ſword in tierce, throwing your body well back in a ſtraight line with that of the adverſary, and ſupporting it on your left ſide.

OF

### *OF THE PASS IN CARTE, AFTER THE FEINT IN CARTE OVER THE ARM.*

IF you are engaged in carte, you muſt make an attack with the foot, and diſengage nicely; holding your wriſt in carte over the arm, make a half thruſt, by advancing your right foot about ſix inches, your wriſt well raiſed, and the point of your ſword firm in a line with your adverſary's face; and at the time he comes to join your blade to parry, you muſt ſurprize him by diſengaging artfully in carte, and make the paſs of the left leg forward, oppoſing your ſword to his, and keeping your wriſt high, and your point in a line to his; this done, recover, and ſeek his blade with a circle parade.

### *OF THE PASS IN CARTE OVER THE ARM, AFTER THE FEINT IN CARTE.*

IF the adverſary is engaged on the outſide of the ſword, you muſt make a half thruſt on the blade with an appel of the foot, and at the time that you feel him join and force your blade, feint on the inſide of the ſword, ſubtilly diſengage on the outſide, making your paſs

paſs with the left leg, and thruſt carte over the arm, according to the rules of this paſs before-mentioned.

You ſhould obſerve, that all the paſſes made on the outſide of the ſword, are parried with a high wriſt in tierce; and thoſe on the inſide are parried by the prime, or a high wriſt in carte, the wriſt being always in a line with the face.

### *AN EVASION, OR SUBTERFUGE, BY THE REMOVAL OF THE LEFT LEG BACKWARD, AT THE TIME THE ADVERSARY MAKES HIS PASS ON THE OUTSIDE OF THE SWORD.*

## PLATE XXXI.

YOU muſt give opening ſufficient on the outſide to encourage your adverſary to make a paſs, either in tierce or carte over the arm; and you muſt not ſtir or move by any attacks of the foot, or half thruſts, which he may make, but ſtill leave an opening to the outſide, to determine him to execute his thruſt, and at the time he makes his paſs with the left foot, drop your point in ſeconde, with the wriſt in tierce, without ſeeking his blade, and carry back your left foot, to its full extent of a ſtraight knee; nor ſtir the right foot, but

*An Evasion or Subterfuge, by the removal of the left Legg backward, at the time the Adversary makes his pass on the outside of the Sword.*

Publish'd as the Act directs Aug.st 1783

keep the right knee bent as on a thruſt, having your body a little lower than in the ſeconde thruſt, your right arm very ſtraight, the wriſt up, and oppoſed to the face, the left arm falling perpendicularly between the two thighs, the palm of the hand open, facing the ground, for fear of a ſlip of the left foot backward, (which would make you loſe your central poſition,) and to enable you to keep up the body, by the aſſiſtance of the hand, in ſuch a caſe. This operation being performed, you muſt recover to a garde, and ſeek the blade by a circle.

### *OF THE SEIZING OF THE SWORD AFTER THE PASS ON THE OUTSIDE OF THE SWORD.*

AFTER having made the paſs, either in tierce or carte over the arm, if the adverſary parries the thruſt, and reſiſts to the blade, you muſt with agility and ſubtleneſs, ſeize the guard of his ſword with your left hand, that inſtant paſſing your right foot before your left, with your heels in a line, bending the knee a little, and ſtraightening the left knee; by this poſition you will be able to oppoſe his defence; if he ſhould take hold of your blade with his

his left hand, you ſhould immediately draw in your right, and preſent your point to his belly, holding faſt the ſhell of his ſword, to oblige him to give it up.

### *OF THE COUNTER DISENGAGE ON THE INSIDE OF THE SWORD, STANDING STILL.*

IF you are engaged within the ſword, you ſhould bear a little on the adverſary's blade, to induce him to diſengage, and at the time of his diſengaging, either in tierce or in carte, to join your blade, you muſt ſeize that moment to time his motion by a counter diſengage, before he touches your blade, and thruſt a well oppoſed thruſt in carte; the thruſt made, recover immediately to the ſword in carte.

### *OF THE COUNTER DISENGAGE ON THE OUTSIDE OF THE SWORD STANDING STILL.*

YOU muſt engage the adverſary on the outſide, with the wriſt turn'd carte over the arm, the wriſt and point being in a line with the ſhoulder, the arm flexible, and you muſt bear on his blade to determine him to diſengage; and at the time he diſengages to join your

 blade

blade on the outſide, diſengage ſubtilly with him before he joins your blade, and thruſt a full carte over the arm, oppoſing your wriſt and ſword according to the rules explained; the thruſt being made, recover quickly to the ſword, and redouble a thruſt in ſeconde.

### *OF THE COUNTER TO THE COUNTER DISENGAGE, STANDING STILL.*

IF the adverſary ſhould offer to parry with a counter diſengage the thruſts abovementioned, you ſhould, without ſeeking his blade, double your diſengage with ſpirit, and ſhunning his blade, thruſt with a ſtraight point at his body, and recover to the ſword by the circle parade.

### *OF THE COUNTER DISEGAGE ON THE TIME, AND OF THE COUNTER TO THE COUNTER, WHEN THE ADVERSARY ADVANCES.*

YOU muſt get out of diſtance about a foot, without leaving the adverſary's ſword; and at the time he advances, ſeize the opportunity, without ſeeking his blade, diſengage, and thruſt ſtraight at him; if he ſhould parry with a counter diſengage, you ſhould redouble the

the diſengage without ſeeking his blade, and thruſt out fully at him, having your body firm, and an oppoſite wriſt, according to the rules mentioned before: the thruſt being made, recover to a guard by a circle.

### *OF A STRAIT TIME THRUST, UPON ANY LOW FEINT.*

IF you are in diſtance with your wriſt turned in carte to the height of the ſhoulder, the point a little lower than the wriſt, and firm on your legs, you ought not by any means to be flurried, or ſtir, by any motion of the adverſary, either by appels or half thruſts; but be watchful to take advantage of any opening he may give at the time he lowers his wriſt and point, and makes low feints; at that inſtant, without ſeeking his blade, thruſt ſtraight forward in carte, with a well oppoſed wriſt, according to the rules: the thruſt made, recover, and ſeek the blade by a circle.

You execute this ſame time thruſt alſo after having retired about a foot, by obſerving, the moment the adverſary lifts up the foot to come in diſtance, to leave his blade, and, if in the leaſt his wriſt ſhould be low, and you ſee opening enough, thruſt ſtraight at him, without ſeeking his blade, or heſitating in the leaſt.

If he ſhould uncover the lower parts of the body at the time he comes in diſtance, you muſt time him, and thruſt a low carte with a well oppoſed and maintained wriſt, and recover to the ſword in tierce.

If you are engaged on the outſide of the ſword, with the wriſt in tierce, you ſhould retire about a foot, and at the time the adverſary advances, giving the leaſt opening, you may thruſt carte over the arm; if he ſhould bear, and force your blade in the advance, you ſhould diſengage, and ſtop him by a carte within the blade.

Nothing is more difficult than to thruſt with exactneſs and preciſion; the abovementioned time thruſt cannot be performed well but by thoſe who have acquired, by practice and experience, theſe quick requiſites of the eye and wriſt to execute with.

To ſhun the conſequences which might ariſe from theſe time thruſts; you ſhould be very attentive to hold your ſword well before you, nor give any openings by the feints you make, nor bear too much on the enemy's ſword in the attacks, either ſtanding ſtill or marching; and be always ready with a counter diſengage, cloſe and quick, or a circle; and never form an idea of giving a thruſt, without that of recovering quickly to a guard.

*THE*

## *THE HALF ROUND, OR BOUNDING TURN OF THE BODY, CALLED DEMI-VOLTE.*

### PLATE XXXII.

WHEN you are in diſtance, you muſt engage the ſword in tierce, having the arm flexible, your body well on the left hip, and give your adverſary opening ſufficient on your outſide to engage him to thruſt; and at the time he forces your blade, either in tierce or carte over the arm, you muſt ſubtilly diſengage your point under the mounting of his ſword, with your wriſt as in carte, well oppoſed, and direct your point to his right breaſt; at the ſame time carry your left foot near the right, that the point of your left foot be about two inches behind the right heel, the two feet forming an angle: to complete this poſition, you ought to keep your knees ſtraight, your head erect and in a line with the right ſhoulder, and the left ſhoulder well turned out, with a ſtretched out arm, as in the carte thruſt.

IT is impoſſible to complete this volte, or turn of the body, with that vivacity and exactneſs it requires, without turning the point of the right foot inward, and lifting it up from the

the ground a little, and turning upon the heel your leg and thigh, as on a pully, that the body may be more thrown back in a ſtraight line with that of the antagoniſt, in order to ſhun the enemy's point; this done, you muſt recover with your left foot back, and come to a guard, according to the rules explained, with the circle parade.

THIS ſame demi-volte may be made at the time the adverſary advances and forces your blade; the complete executing of this thruſt conſiſts in taking the juſt time.

## *OF THE FLANCONADE THRUST, HAVING PARRIED THIS BOUNDING TURN, CALLED DEMI-VOLTE.*

YOU muſt force the adverſary's blade on the outſide, to engage him to thruſt within; and if he ſhould make this demi-volte, you ſhould parry the thruſt, with the fort on his feeble, and binding the ſword, return a flanconade, with the oppoſition of the left hand, as before explained; this done, recover your ſword in carte.

*THE*

*The Half Round or bounding turn of the Body calld Demy Volte.*

*Publish'd as the Act directs Aug.t 1783*

### *THE WHOLE ROUND OR TURN CALLED THE VOLTE, AT THE TIME THE ADVERSARY DISENGAGES TO THRUST IN CARTE.*

AT the time the adverſary diſengages within, and thruſts carte, you muſt raiſe the wriſt to the height of your face, with your nails upward, and with ſwiftneſs and agility perform this turn of the body called volte; fixing your point to his right breaſt, ſtraighten your legs at the time you paſs with the left foot behind the right, which diſtance ſhould be about a foot; your left ſhoulder ſhould be turned outward, to form a complete ſide front or profile, to your adverſary, your head in a line with your ſhoulder and arm, to enable you to direct your point forward in that line. The thruſt being made, recover your guard with your wriſt in tierce, and beat ſmartly, with the fort of your tierce edge,[†] on the feeble of his ſword, raiſing your wriſt and lowering your point.

*The whole turn call'd Volte, on the Pass made in Tierce, or in Carte over the Arm.*

*Publish'd as the Act directs Aug. 1783*

## *OF THE WHOLE TURN CALLED VOLTE, ON THE PASS MADE IN TIERCE, OR IN CARTE OVER THE ARM.*

## PLATE XXXIII.

AT the time the adverſary makes the paſs on the outſide of the ſword, you muſt paſs your point under his arm, and turn your wriſt in carte to the height of the face, paſs the left foot behind the right, about a foot diſtance, and fix your point to his right breaſt, ſtraightening both legs; at the very time you paſs the left foot, be careful that your left ſhoulder be well turned out, ſo as to be with your back half turned to the adverſary, holding your head in a line to the right ſhoulder and arm, in order to carry the point directly forward: this thruſt being executed, recover your guard by a circle, and keep your body very ſtaunch on the left part of the body.

*A Disarm after having parried the Carte Thrust.*

*Publish'd as the Act directs Aug.t 1783*

## *OF A DISARM, AFTER HAVING PARRIED THE CARTE THRUST.*

## PLATE XXXIV.

IF the adverſary is irregular and careleſs when he thruſts a carte, you ſhould parry him with the carte parade, by a dry, ſmart beat with your fort, at the ſame time advancing your right foot about a foot, and ſtraightening your left leg, you muſt ſeize the ſhell of his ſword, with your left hand, and holding it faſt, preſent your point to his body under his arm; if he ſhould make any reſiſtance, and not ſurrender his ſword, you ſhould immediately bring up your left leg to the right, and with the fort of your ſword bear ſtrong on his blade, which will oblige him to open his fingers, and drawing in your arm, ſtill holding his ſword faſt, you will become maſter thereof: the diſarm being made, carry your left foot two feet back, with a ſtraight knee, and preſent the the two points at him, as you ſee in plate 35th.

2d. Position of the Disarm, after having parried the Carte Thrust. —

Publish'd as the Act directs Augt. 1783

*A Disarm on the Thrust in Tierce, or Carte over the Arm.*

Publish'd as the Act directs Aug.st 1783

## *OF THE DISARM ON THE THRUST IN TIERCE, OR CARTE OVER THE ARM.*

### PLATE XXXVI.

IF the adverſary makes a thruſt in tierce, or carte over the arm, and abandons his body in a careleſs manner, you muſt parry him by a dry ſmart beat with the edge of your fort, traverſing the line of the blade, and force or bear his wriſt upwards,[†] at the ſame time paſſing the left foot about a foot before the right; if he ſhould reſiſt, or bring up his left foot to cloſe in and ſeize your blade; in ſuch a caſe, ſtill holding faſt his ſword, you ſhould throw his arm outward to the right,[†] and carry your left foot forward about two feet, bending your right knee a little, and ſtraightening the left, preſent the point of your ſword to his face, raiſing your wriſt and arm to the height of your face, as you may obſerve in plate 37th.

*The Second Position of the Disarm after having parried the Thrust in Tierce.*

*Publish'd as the Act directs Aug.t 1783*

*The disarm on the Carte or Second Thrust, after having parried with the Prime Parade.*

*Publish'd as the Act directs Aug.t 1783*

### *OF THE DISARM ON THE CARTE OR SECONDE THRUST, AFTER HAVING PARRIED WITH THE PRIME PARADE.*

## PLATE XXXVIII.

IF you are engaged in tierce, make an attack of the foot, and force the enemy's blade on the outſide, to excite him to thruſt; and at the time he thruſts either carte or ſeconde, parry quickly with the prime; and inſtead of traverſing the line to the right, as I have before mentioned in the articles of the parades, you muſt advance about half a foot, and with ſwiftneſs paſs your right arm† over the fort of his blade; by this means, by drawing in your body and your left arm,† he will be forced to quit his ſword: as ſoon as the diſarm is made, preſent your point, and paſs ſwiftly back, with your right foot at a foot diſtant from the left, as you will ſee in plate 39th.

*Second Position of the Disarm on the Carte or Second Thrust, after having parried with the prime Parade.*

Publish'd as the Act directs Aug.t 1783

## *OF THE DISARM AFTER THE PARADE ON THE OUTSIDE OF THE SWORD.*

### PLATE XL.

IF you are engaged on the oútſide, either in tierce or carte over the arm, you muſt make an appel of the foot, and force or bear a little on his blade, to excite him to thruſt a carte within the ſword.

AT the time he diſengages and thruſts out, you muſt counter-diſengage and parry, forcing his blade upward with the fort of yours: you are to paſs your left foot before the right, about the diſtance of a foot, and with livelineſs and reſolution, with your left hand, ſeize the ſhell of his ſword; and as in defending himſelf he might bring up his left leg, and throw himſelf forward on the blade; to hinder his ſeizing it, you ſhould inſtantly throw your right ſhoulder and arm back, and carry your right foot behind the left about a foot, and turning the point of your left foot facing his knee, and paſſing your ſword behind your back, leaning your wriſt againſt your loins, preſent the point of your ſword to his belly.

You

*The Disarm after the Parade on the outside of the Sword*

*Publish'd as the Act directs Aug.t 1783*

You muſt always obſerve that, in all diſarms, you are never to ſeize the adverſary's arm, nor his blade, with your left hand; for in ſeizing his arm he may again recover his ſword, by ſhifting or throwing it from the right to the left, and having caught hold of it by the gripe, or by the fort of his blade, he may with a drawn in arm ſtab you; and was you to ſeize his blade, he might alſo draw in his arm, and draw it through your hand in a fatal manner. In my opinion, all thoſe diſarms which I have explained, are more brilliant and fine in a fencing-ſchool, with a foil in hand, when very well executed, with the utmoſt precision and judgement, than they are uſeful ſword in hand; nevertheleſs, they may be made uſe of againſt thoſe who abandon their bodies after they thruſt, and who do not recover with that quickneſs and care which is neceſſary. Sword in hand, I prefer the bindings and croſſings of the blade, or the ſmart dry beat with the fort on the feeble of the adverſary; by all theſe you run no riſque; for if you don't beat the ſword out of the hand, you will always get opening thereby to throw in a thruſt if you deſign it.

*OBSERVATIONS*

## *OBSERVATIONS ON LEFT HANDED FENCERS.*

IT often happens that the right handed fencer is much embarraſſed in defending himſelf againſt a left handed one, occaſioned by the conſtant habit of fencing always with right handed fencers, which gives the left handed fencer a conſiderable advantage. You ſeldom have occaſion to fence with a left handed man, becauſe the number of theſe is but ſmall; and for the ſame reaſon, when two left hands meet, they are equally at a loſs with one another.

To obviate this inconveniency, I am of opinion, that a fencing maſter ſhould accuſtom his ſcholars to fence with both hands; (that is to ſay) that when the pupil has learnt to handle his foil well with the right hand, he ſhould be exercifed with the left hand. This practice will be found hard to every body, but with a good will, and by taking pains, you may attain to a degree of perfection which will be advantageous to yourſelf, and will do honour to him that teaches.

THE maſter ſhould not only uſe his ſcholars to take leſſons with both hands, but ſhould likewiſe uſe them to fence looſe, called aſſaulting; this method would enable them to defend themſelves

themſelves with both hands, and they would never be at a loſs againſt an adverſary who might preſent himſelf to them in a different poſition than their own.

WHEN a right handed and a left handed fencer are together, they ought to be attentive, both of them, to keep the outſide of the ſword; this ſide being the weakeſt, they have both of them the facility of beating, or making a glizade or preſs on the outſide of the blade.

IF the beat is given properly, it is almoſt impoſſible that the ſword doth not drop out of the hand, except the adverſary takes the preciſe time of the beat, either by diſengaging, or by turning his wriſt in tierce.

YOU muſt obſerve alſo, that the right handed fencer ought to thruſt carte inſtead of tierce, to the left handed one, and tierce inſtead of carte; that is to ſay, that he ought to thruſt all the outward thruſts within, and the inner thruſts without.†

THE ſame rules alſo are for the left hand to the right handed fencer; by this means the hand will always be oppoſed to the ſword, and the body and face will always be covered.

## *OBSERVATION ON THE GERMAN GUARD† OF THE SMALL SWORD.*

IN the pofition of the German guard the wrift is commonly turned in tierce, the wrift and arm in a line with the fhoulder, the point at the adverfary's waift, the right hip extremely reverfed from the line, the body forward, the right knee bent, and the left exceedingly ftraight. The Germans feek the fword always in prime or feconde, and often thruft in that pofition with a drawn in arm. They keep their left hand to the breaft, with an intent to parry with it; and the moment they draw their fword they endeavour to beat fiercely with the edge of their fword on their antagonift's blade, with an intent to difarm them if it be poffible.

## *THE DEFENCE AGAINST THE GERMAN GUARD.*

IN order to vanquifh this guard, you muft prefent yourfelf out of diftance, and brifkly attack with a beat of the foot, and make a half thruft on the infide, towards the face of the adverfary, to oblige him to raife his wrift; in that time difengage over the fort of his fword, and

and thruſt a ſecond thruſt at full length. You may alſo put yourſelf in guard in the ſeconde poſition, keeping your point directly in a line to his arm-pit, and feint from an outſide to an inſide over his blade, and at that time make an appel of the foot, to oblige him more eagerly to come to the ſword; then ſeize the exact time to make a diſengage over his blade, and thruſt a ſeconde or a quinte thruſt. If he parries this thruſt you ought to triple your diſengages, and hit him in prime or carte over the arm.

If you are diſpoſed to wait the attack of the adverſary, you muſt put yourſelf in guard with a high tierce, and your point fixed at the adverſary's right ſhoulder, and not move or flutter by any motions he can make; except he is ſufficiently out of diſtance to make a ſtraight time thruſt. If he ſhould keep ſtaunch on his guard, you ſhould give him opening ſufficient on the inſide to encourage him to thruſt in there; and if he does, you ſhould parry with prime or half circle, traverſing the line on the right,† and returning the thruſt with ſpirit, keeping your wriſt in the ſame ſituation of the thruſt which you parried: if he doth not attack, or is not moved at any attacks made on him, you muſt place yourſelf in poſition of a high tierce, as I have before explained, turning with ſubtilty your wriſt from tierce in carte, ſlipping your point over the fort of his blade, which will form a demi-circle, beat ſtrongly with the fort of your inſide edge on his blade, and immediately thruſt a full ſtretched out carte; by this means it will not be very difficult to throw his ſword out of his hand.

If he ſhould parry with the left hand, obſerve never to thruſt within the ſword till you have made a half thruſt, well maintained with the wriſt, to baulk his left handed parade.

## *EXPLANATION OF THE ITALIAN GUARD.*†

THE Italian guard is commonly very low; they bend equally both knees, carry the body between both legs; they keep the wriſt and point of the ſword low, and have a contracted arm; they keep the left hand at the breaſt, to parry with it, and ſtraightway return the thruſt.

Though this guard is natural to them, yet they vary every moment, to perplex their adverſaries, in keeping a high wriſt, and point to the line of the ſhoulder; in keeping a high wriſt and a very low point, and making large geſticulations of the body, and turning round their antagoniſt, ſometimes to the right, and ſometimes to the left, or by an immediate advance of the left foot to the right; and they thruſt ſtraight thruſts at random, or make paſſes and voltes: they have much dependence on their agility, and the parade of the left hand; for

for that reafon, when two Italians fight together they often are both hit together, which is called a counter thruft: this happens feldom with two good fwordfmen, becaufe they know how to find the blade by a counter difengage, or by the circle, and becaufe they have a quick return.

AND yet, neverthelefs, I am perfuaded that the above Italian method would puzzle a good fwordfman, if he did not take the neceffary precautions which I am going to explain in the next chapter.

## *THE DEFENCE AGAINST THE ITALIAN GUARD.*

IN order to defend yourfelf againft this Italian method, you ought to be very cool, and put yourfelf in a pofition quite covered, and never ftir at any of thefe different motions.

You fhould attack frequently, make half thrufts out of diftance, to entice him to clofe in, and at the time he lifts up his foot to come in diftance, execute your thruft, without ftirring your wrift or your fword from the line of his body; that if in cafe he had intended this advance as a thruft, you may be thereby enabled to parry, and return the thruft immediately.

You ſhould never redouble with ſuch people, for fear of a counter thruſt and the parade of the left hand, which would occaſion a return; and you ſhould at all times, after the delivery of a thruſt, whether you hit or not, recover immediately to a guard with the circle parade.

If the Italian ſhould ſtand before you with his arm and his point in a direct line, you ſhould make uſe of the binding of the blade, or of the beats, and thruſt ſtraight and firm at him. You ſhould never be fond to thruſt to the great openings he may give, for fear of a time thruſt, but make a half thruſt, and if he ſhould thruſt at that time, you muſt parry, and cloſe in about ſix inches, and with ſpirit and reſolution return the thruſt.

To baulk the parade of the left hand, you muſt execute a half thruſt, and finiſh it the moment the motion of the left hand is made, in order to parry therewith.

*N. B.* This is only good to put in practice againſt thoſe who are not fond of returning a thruſt.

You ſhould alſo never be fond of thruſting to the adverſary's inſide when he gives a large opening; but you may feint on the inſide, and thruſt on the outſide, or the lower part of the body.

If

If he ſhould bring up his left foot to his right, you ought to make a beat on his blade, or deliver a half tnruſt; and if by this he moves not, you ſhould ſwiftly get out of diſtance, by carrying your right foot up to your left, parrying at the ſame time with the circle parade, or wait till he thruits, which if he does, you muſt ſeek his blade by a counter diſengage, and either ſeize his ſword, or return a thruſt the moment he makes his retreat.

## *OF THE ITALIAN GUARDS WITH THE SWORD AND DAGGER.*

## PLATE XLI.

THIS exerciſe of ſword and dagger is only made uſe of in Italy. When the ſcholar has learnt to handle his ſword well he is afterwards inſtructed how to uſe the ſword and dagger: the Italians ſeldom go out at night without theſe two weapons. The right handed man carries his dagger by the ſide of his right thigh, and the left handed man by his left: they draw this weapon the moment they have ſword in hand. Naples is the city where theſe are moſt commonly uſed, and with moſt dexterity.

THE dagger is never made uſe of in Paris, but at the public reception of a fencing maſter: when an uſher has finiſhed his apprenticeſhip under an able maſter, and is preſented to the public to be received as a maſter, he is obliged to fence with ſeveral maſters. After having performed with the foil alone, he is to fence with ſword and dagger. The reception of a fencing maſter hath ſomething pleaſing in it, and gives the more emulation to youth to be inſtructed in that art, ſince no man can be received among the maſters unleſs he hath ſerved a regular ſix years apprenticeſhip under one maſter (a cuſtom only made uſe of in Paris). This public exerciſe, or trial, which is as the touchſtone of the art of the ſword, called fencing, produces an effect the more advantageous, as it tends to the perfecting of that art.

I SHOULD be ungrateful if I was ſilent on the ſuperior talents of the French fencing maſters; and, according to the knowledge which I have acquired, I believe them to be the beſt in the world, both for their graceful attitudes and profoundneſs of knowledge.

THOUGH there is no uſe made of the ſword and dagger in this country, I thought it neceſſary to give an explanation thereof, that gentlemen may know how to defend themſelves if they ſhould travel in countries where they are uſed, and not be embarraſſed when they ſee two points at once before them.

I SHALL

*The Italian Guards with the Sword & Dagger.*

*Publish'd as the Act directs Aug.t 1783*

I SHALL therefore give two different guards, which are the moſt made uſe of in this exerciſe; and will afterwards explain the manner in which a ſingle ſword is to defend itſelf againſt the ſword and dagger.

YOU muſt place yourſelf in guard, with the dagger and arm ſtretched out, and at ſome diſtance from the hilt of your ſword, to execute and form the parades as cloſe as poſſible; which is very difficult with a ſtraight arm. You muſt obſerve that, in covering one ſide, you do not uncover another. In this guard there is no ſingling the body, for the left ſhoulder projects more than the right; and though the right arm covers the outſide of the ſword, it ought to be contracted.

WHEN in this poſition you can form your parades well, you will put yourſelf in guard, your ſword arm ſtraight but not ſtiff, and your left drawn in, having the point of your dagger near the right elbow.

THE principal point is not to flutter, or ſtir at any motion made by the adverſary; if he ſeeks your ſword with his, you muſt ſlip him, unleſs you find yourſelf firm enough to oppoſe him therewith, cloſing in about ſix inches, and without quitting his ſword, ſtrive to get his feeble on the fort of your dagger, and quitting his ſword, it will be eaſy to deceive his dagger,

*The return in Tierce after having parried with the Poignard.*

*Publish'd as the Act directs Aug.t 1783.*

ger, and hit him. In this operation, you muſt not quit his blade with your dagger, and the longer his ſword is, the greater will be your advantage, and alſo on any parade made with the dagger, you ought not to quit the blade, if you have a mind to return the thruſt.

THE Italians frequently parry with the dagger, therefore it is evident, that he who can parry with two blades has a great advantage, provided it be done without hurry, and with judgment, for otherwiſe he would only leave himſelf continually open.

YOU ſhould baulk your adverſary at the time he makes an attack, or half thruſts, by not ſtirring your ſword, but make ſome wide motions with the dagger, to engage him to thruſt, and as ſoon as he delivers his thruſt, parry with your dagger; cloſe at the ſame time in with him, make a feint with your ſword toward his face, and thruſt at the body, as you ſee in plate 42.

THE Italians defend all the inſide, and the lower part of the body, with the dagger, and as they depend entirely on this parade, they lower the outſide with the point of their ſwords.

## *OF THE SINGLE SWORD, AGAINST THE SWORD AND DAGGER.*

AS all the thrufts which the fingle fword makes on the infide, againft the fword and dagger, may eafily be parried, the return of the adverfary's point would infallibly hit, therefore you fhould act with great judgment and attention.

You muft come to a garde, with your wrift turned between tierce and carte, and a little lower than the ordinary garde, fixing the point to the adverfary's right fhoulder, you muft not engage his fword, but make frequent beats on his outfide, and attacks with the foot, always directing your point to his face, to oblige him to raife his wrift, which time you muft feize with precifion, and with fwiftnefs and vivacity, deliver your thruft in feconde. and return as quick to the prime parry or circle.

If he fhould be in guard with the point in a line with his fhoulder, you fhould feint on his infide, and return with a beat on the outfide, from your fort on his feeble, and deliver a thruft carte over the arm.

If

If he ſhould hold his point lower than his wriſt, you ſhould place yourſelf ſo likewiſe on his inſide, and making a half thruſt on the inſide, immediately bind his blade ſmartly, and thruſt a flanconade. You may alſo after the half thruſt, croſs bind his ſword, and make your thruſt in tierce; I would not adviſe any body to thruſt on the inſide, becauſe the dagger will be very apt to parry, and thereby you would be liable to the return of the ſword, but when the adverſary is not ſtaunch in his parades, and flutters, ſeeking to parry with his ſword the attack you make on him, in ſuch a caſe, after having made a half thruſt on the inſide, and on the outſide of the dagger, you may deliver a low thruſt in carte; the thruſt made, recover to a guard in tierce, or an half circle.

### *OF THE SPANISH GUARD*† *MARKED* A, *ATTACKED BY THE FRENCH GUARD.*

### PLATE XLIII.

THE Spaniards have in fencing a different method to all other nations; they are fond often to give a cut on the head, and immediately after deliver a thruſt between the eyes and the throat. Their guard is almoſt ſtraight, their longe† very ſmall; when they come in

diſtance

*The Spanish Guard Mark'd A. attack'd by the French Guard.*

Publish'd as the Act directs, Aug.^t 1783

diſtance they bend the right knee and ſtraighten the left, and carry the body forward; when they retire, they bend the left knee and ſtraighten the right, they throw the body back well, in a ſtraight line with that of the antagoniſt, and parry with the left hand, or ſlip the right foot behind the left.

THEIR ſwords are near five feet long from hilt to point,† and cut with both edges; the ſhell is very large, and behind it is croſſed with a ſmall bar, which comes out about two inches on each ſide; they make uſe of this to wrench the ſword out of the adverſary's hand, by binding or croſſing his blade with it, eſpecially when they fight againſt a long ſword; but it would be very difficult for them to execute this againſt a ſhort ſword. Their ordinary guard is with their wriſt in tierce, and the point in a line with the face. They make appels or attacks of the foot, and alſo half thruſts to the face, keep their bodies back, and form a circle with the point of their ſwords to the left, and ſtraightening their arm, they advance their body to give the blow on the head, and recover inſtantly to their guard, quite ſtraight, with their point in a direct line to their adverſary's face.

OF

### *OF THE SPANISH GUARD DEFEATED, AFTER THE ATTEMPT OF THE CUT ON THE HEAD.*

## PLATE XLIV.

IF you make uſe of a ſword of common length, and if you can but ſtand the firſt attack, you will eaſily defend yourſelf againſt a Spaniard, and will be very little embarraſſed by his play or method.

You ought to put yourſelf in guard out of diſtance, with your wriſt turned in tierce, holding it a little higher than in the ordinary guard, with great coolneſs, nor anſwer any motion he may make or attempt. If he ſhould attempt the cut on the head, you ſhould parry it with a high tierce, ſtill raiſing your wriſt and bending your body, and cloſe in about a foot or more ; after which briſkly return a full ſtretched out thruſt in ſeconde, with your point lower than common in that thruſt, that he may not be able to parry it with his left hand. The thruſt being made, recover inſtantly to a tierce, and traverſing the line to the right, with a forcibly oppoſed wriſt ſeek his ſword again ; at the ſame time bring up your right

*The Spanish Guard - defeated, after the attempt of the Cut on the Head.*

Publish'd as the Act directs Aug.t 1783

right foot to your left, to throw off his point: ſo will you be enabled to get ground to advantage with the left foot. If the adverſary makes a thruſt to the face or body, you muſt parry it by diſengaging from tierce to carte, keeping your wriſt in a line with the ſhoulder, and at the ſame time cloſe in a full foot, to get within his blade as much as poſſible, and to be able to return a thruſt in carte. If he wants to parry with his left hand, you muſt feint on it, making at the ſame time an appel of the foot, to baulk his left handed parade, and finiſh your thruſt according to the forementioned rules; recover quickly your ſword, with your point to his face, and redouble a low carte.. This done, recover to a guard, carrying your right foot behind your left.

THOUGH it ſeems eaſy for the ſhort ſword to diſarm the long ſword after you have the advantage of getting within his blade, I would neverthleſs adviſe nobody to attempt it, for fear you ſhould not be able to reach the ſhell of his ſword, or for fear of having your fingers cut by the edges of it.

I ALSO would not adviſe any body to croſs or bind, or to beat on their blades, becauſe the Spaniards, when they draw their ſwords, paſs the two firſt fingers through two ſmall rings which are near the ſhell, and with the two others and the thumb they have a faſt hold of their

their gripe: therefore it is evident that none of these last mentioned operations would be succesſsful.

*THE POSITION OF THE GUARD CALLED SWORD AND CLOAK, BY THE SWORD AND LANTHORN.*

## PLATE XLV.

THE ſword and cloak, which is an ancient cuſtom made uſe of in Italy, has never been forbidden by the government, as has the ſword and dagger in various places of that country. The cloak is offenſive and defenſive: it is offenſive, becauſe thoſe that are very expert in the uſe of it, have it in their power to be hurtful to their adverſary. There are many ways to throw it; you may not only cover the whole ſight of the enemy, but his ſword alſo: but if to the contrary, a man is not expert in it, he may cover his own ſight, and obſtruct his own ſword alſo, and therefore be a prey to the adverſary. It is defenſive, becauſe it obſtructs all the cuts that may be made to the head or body, if given within the ſword; the outſide blows, either over or under the hilt, ſhould be parried with the ſword, and

*The Guard of the Sword and Cloak, opposd, by the Sword & Lanthorn.*

*Publish'd as the Act directs Aug.t 1783*

and the ſword ſhould be ſeconded or aſſiſted by the cloak, that the ſword may return the cut and thruſt.

To make the proper uſe of the cloak, you ſhould wrap your left arm round with part of it, and let drop or hang down the other part, but take care it hangs no lower than the knee; and obſerve, if you were obliged, after a long defence, to drop the arm a little, to reſt it, not to drop the cloak to the ground, or before your feet, for fear of treading on it, and thereby getting a fall.

If you find yourſelf fatigued with the left arm, you may reſt it by dropping it along your ſide, keeping the cloak at a little diſtance from your thigh, and making a paſs backward; ſoon after recover to a guard. If you ſhould not have room to retire, you may lean your left hand on your hip, and keep your ſword in a continual circle parade.

It is very eaſy to a perſon who underſtands the ſword and dagger to make uſe of the cloak, becauſe this defence requires a quick and exact ſight. In caſe of need, one might defend one's ſelf againſt a ſword with a cane and cloak; for after having parried a thruſt of the ſword with a cane, one ſhould cloſe in at the ſame time, without quitting his blade, and cover his



head

head with the cloak. To perform this operation well, one ought to be well ſkilled in fencing, very cool and reſolute.

One ought alſo well to underſtand diſtance to uſe the cloak; and to execute a deſign well, one ought to give an opening to the adverſary, to engage him to thruſt, and immediately, without in the leaſt ſeeking his blade with the ſword, throw the body backward, and fling the hanging part of the cloak againſt his ſword; and traverſing from the ſtraight line, return a full thruſt with your wriſt in carte.

### *EXPLANATION OF THE GUARD CALLED SWORD AND DARK LANTHORN.*

THOUGH there are ſevere puniſhments inflicted on thoſe who are found ſword in hand with a dark lanthorn, yet there are ſome to be met with from time to time; therefore I think it neceſſary to ſhew the manner of defence againſt it. Thoſe who uſe the dark lanthorn commonly hide it under their clothes or cloak; and when they attack any body they open it before they draw their ſword, and preſent it before them either above their head, or behind them, by turning the hand behind their back; and change the poſition thereof

as

*The Guard of the Sword & Lanthorn, opposd by the Sword & Cloak.*

*Publish'd as the Act directs Aug.t 1783*

as the adverſary changes his poſition. If they hold their lanthorn before them, and one is provided with a good ſword, one ought to cover the inſide well with the cloak, and give a ſmart beat on the inſide of their blade, and redouble it with a back handed blow in tierce on the wriſt which holds the lanthorn :† this blow ought to come from the right to the left, and it ought to be executed from the half arm only to the wriſt, that the whole arm go not aſtray too much by it, and that one may be able to return a thruſt with the wriſt in the ſame ſituation, covering the inſide of the body with the cloak.

If he ſhould preſent the lanthorn over his head, you ſhould traverſe to the right,† and get the advantage of the outſide of his ſword, making half thruſts to the face. If he raiſes his point you ſhould cloſe in diſtance, holding both hands high, and keeping the blade over the left wriſt and cloak, and make a paſs with the left foot without leaving his ſword, and aſſiſting with the cloak, draw in the right arm a little, to diſentangle your point, and in the poſition your wriſt finds itſelf ſituated at that time, thruſt with ſpirit and agility directly at the adverſary.

If he preſents the lanthorn by the ſide, with his arm turned behind him, you muſt traverſe the line to the inſide, holding your hand and cloak in the line with your right breaſt,



turning

turning your right hand in tierce, the point of your ſword directly to his belly ; and the moment he delivers his thruſt, inſtead of parrying it with your ſword, ſtretch out your left arm and cover his blade with your cloak, at the ſame time thruſting at the body, as you ſee in plate 46th.

If the adverſary ſhould be garniſhed (that is, ſtuffed with ſomething within his clothes, to prevent a thruſt going through to the body) which you will find out by the thruſt being planted at his body without effect ; in ſuch a caſe, you muſt thruſt at the throat or at the face, or at the lower part of the waiſt ; for it is ſuppoſed that the man who will make uſe of ſo unlawful means as a dark lanthorn in any particular combats, will not ſcruple to uſe the means of garniſh, as before mentioned.

## *OBSERVATION ON THE USE OF THE BROAD SWORD.*

THE broad ſword has four principal cuts in its play ; which are, at the head, at the wriſt, at the belly, and at the ham ſtrings. Some make their cuts from a motion of the ſhoulder, the elbow or the wriſt ; thoſe keep a ſtraight arm,[†] and preſent the point of their ſwords continually to their adverſary.

THE firſt manner of cutting, from the ſhoulder, is done by raiſing the arm, and making a large circle with the ſword, to gather ſtrength to give the blow. This way of executing, which is the worſt of all becauſe it is the ſloweſt, gives a great advantage to him that points; for if he is attentive to cloſe in at the time the broad ſword raiſes his arm, he may give him a time thruſt, or by ſlipping the broad ſword, and at the ſame time cloſing in and ſingling his body, he may furniſh a timely thruſt. It is plain, that if the broad ſword finds no appuy,† or reſt, either on the body or ſword of him that points, that the blow given in vain will quite throw his blade behind him; or if he cuts downward, it will come to the ground, and may break his ſword; but if either of theſe chances ſhould not happen, his motions are ſo coarſe and ſlow, that it is impoſſible for the point, with the leaſt attention, not to find an opportunity of throwing in a thruſt.

THE ſecond way of cutting, by a motion of the elbow, is by drawing the elbow very much in, and this throws alſo the wriſt much out of the proper line, both under and over, and gives likewiſe a great advantage to the point, though not ſo much as the firſt mentioned, becauſe the motion not being ſo wide, it is quicker, and covers the body more.

THE third is from the motion of the wrist, either from the sword forming its circle from right to left, or the contrary; the wrist ought to act with more swiftness, because the elbow and arm are not thrown out of a line of the body. The broad sword commonly parries the thrusts with the fort of his blade, and returns an edge blow from the wrist; and all his favourite blows are on the outside of the sword.

I HAVE herein placed a guard of the broad sword, marked A, and the small sword guard on the defensive, marked B, as you see in plate 47th.

THE broad sword attitudes, or position for their guard, are various; some keep their wrist turned in tierce, with a straight arm, and their point in a line of their adversary's face, keeping the body somewhat forward, the left knee straight, and the right bent.

THERE are some who keep the fort of their broad sword in a line to their left hip, with a high point.

THERE are others, who keep the hanging guard, called the St. George; and others who bend their left knee, and keep their body back, with their wrist turned in carte.

EXPLA-

## *EXPLANATION OF THE DEFENSIVE GUARD OF THE SMALL SWORD AGAINST THE BROAD SWORD.*

THE guard of the ſmall ſword marked B, againſt the broad ſword marked A, which I have placed here, is the moſt ſafe, and the moſt ſheltered guard for defence. The chief point will be, to know your diſtance: in whatever poſition the broad ſword man may put himſelf, you muſt place yourſelf out of diſtance, and bring neither your wriſt nor your ſword, nor your right foot forward: but the moment you draw your ſword, you muſt, with your left hand, take up the ſkirts of your coat, keeping your left hand to the height of your ear, in order occaſionally to parry the cuts of the broad ſword on the inſide, either at the head, face or the lower part of the body.

THE blow at the head may likewiſe be parried with the fort of your blade, having the wriſt in tierce, and oppoſing the blade almoſt croſſing the line; but your point ſhould be a little higher than the mounting of your ſword: the moment the blow is parried, you muſt cloſe in about a foot, and bending the body a little, return a thruſt in ſeconde, and redouble the thruſt before you recover your guard.

*The defensive Guard, Fig. B. of the Small-Sword against the Broad Sword Fig. A.*

*Publish'd as the Act directs Aug.t 1783*

Parry the cut on the outſide of the blade to the face with the fort of your blade, and your wriſt turned half way to tierce with a ſtraight point. The blow being parried, you muſt return a thruſt to the face in carte over the arm, and redouble the ſame with a ſeconde. The cut at the belly on the outſide of the ſword, is to be parried by turning your wriſt to a ſeconde, and returning on the ſame ſide.

If you parry the inſide cuts which may be made at you with your blade, you ſhould parry them with the prime parade, at the ſame time traverſing the line to the outſide, and return a thruſt in prime.

The ſafeſt and ſureſt defence againſt the broad ſword (in my opinion) is not to be fluttered or moved at any motion, ſham blows, or attacks, which the adverſary may make to intimidate you, but ſlip and ſhun his blows, by throwing back your body well in a ſtraight line with his, and retiring about a foot at a time, and counteract his deſigns by continual half thruſts and appels. If his motions are cloſe, you muſt be the quicker to parry, either with the ſword, or with the ſkirts of your coat, and on occaſion make uſe of both.

If his motions are wide, you muſt reſolve to cloſe in, covering yourſelf as much as poſſible with your ſword and the lap of your coat, and deliver your thruſt where you ſee an opening to to hit him. If the ground ſhould not be level enough to tire him, you ſhould, by turning to the right or to the left, and by retreating, take a favourable and exact juſt time for thruſting, inſtead of throwing the thruſt at random.

There are ſome broad ſword men who intermix their play with thruſts (which is called counter point) they feign to give a cut, and finiſh it a thruſt; and ſometimes, after having parried, according to the opening they find, they will return either a cut or a thruſt.

The ſword called cut and thruſt is very different from the broad ſword, becauſe it is much lighter, it carries a ſtraight point, and not a raiſed one, as the ſabres or cutting ſwords commonly have; for which reaſon they are obliged to make the hilts heavy, to render the point light.

The half cut and thruſt ſword is preferable to the broad ſword, provided it be made uſe of with judgement. This weapon is the beſt for horſemen, when they charge their enemy ſword in hand.

THE END.

# TABLE OR INDEX.



Poſition

## TABLE OR INDEX.



Of

# TABLE OR INDEX.



Of

## TABLE OR INDEX.



Of

## TABLE OR INDEX.



Of

## TABLE OR INDEX.



Of

# TABLE OR INDEX.



Of

## TABLE or INDEX.

# Appendix A

## The Construction of the Small Sword and Its Parts

A Pommel nut

B Pommel

C Ferrule (Turk's head) on grip

D Grip

E Knuckle guard (knuckle bow)

F Grip washer

G Quillon connected to quillon block or cross

H Arms of the hilt

I Ricasso washer

J Shell guard

K Ricasso

L Tang of the blade

M Heel or shoulders of the blade

N Upper section of the furniture of the scabbard, mounted with a ring

O Upper section of the furniture of the scabbard, mounted with a hook

P Upper section of the furniture of the scabbard, mounted with a button

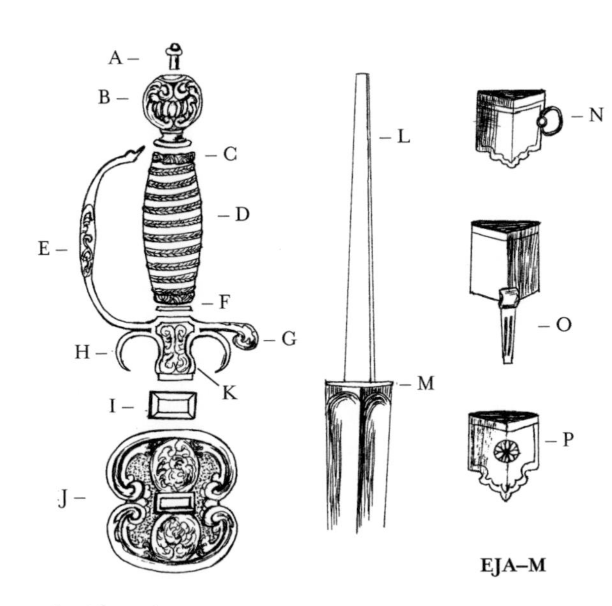

# Appendix B

## Clarifications of Technical Details Omitted by Angelo

There are technical aspects of fencing that are often left out of treatises. These texts are frequently written for the students of a particular master, and a certain amount of technical knowledge is assumed on the part of the reader. Angelo's text is among the most complete, which is one of the reasons it was chosen for Diderot's *Enclyclopédie*. Angelo nevertheless leaves out some key technical points. To assist the reader I have included information from other period sources to clarify these points.

Small swords from the second half of the seventeenth century vary considerably from the small swords of the eighteenth century, yet the two most common forms of gripping the weapon remain the same; these are described below.

In the first method of holding the weapon, the thumb is placed on the cross (where the 'cross bar' and the ricasso meet), with the thumbnail upward and the thumb tip facing the guard. The index finger is slightly extended so that it lies under the flat of the vestigial ricasso. The remaining three fingers are wrapped around the grip, which rests in the hollow space located in the centre of the heel of the hand; the pommel is held on the centre of the wrist. This is described on p. 3 of Danet's 1766 treatise *L'Art des armes*:

> *Pour tenir avantageusement l'épée, il faut que la poignée se trouve entre le tenar & l'hypotenar, & le pommeau à la naissance de la main; que le pouce soit allongé jusqu'à la distance d'environ douze lignes de la coquille sur le plat de la poignée; qu'en même temps le milieu de l'index se place dessous la poignée près de la coquille; que la poignée soit étroitement embrassée par le doigt majeur, & encore ferrée contre le tenar vers le pommeau, par l'annulaire & l'auriculaire: mais il ne faut ferrer la poignée que dans l'instant seulement que vous tirez, ou que vous parez; parceque les muscles du pouce, de l'index & du doigt majeur s'engourdissent promptement, au lieu qu'il n'en est pas de meme de ceux qui sont agir le petit doight & l'annulaire.*

*Il est des occasions où il convient de lâcher ces deux doigts pour faciliter l'exécution de certains coups. J'aurai attention de vous en prévenir quand il le faudra.*

In order to hold the sword advantageously it is necessary that the grip is found between the thenar and the hyopthenar [that is, the muscles controlling the thumb and little finger] and the pommel at the beginning of the hand; that the thumb is stretched out just to the distance of about twelve lines from the shell on the flat of the grip [hilt];[1] at the same time the middle of the index finger is placed below the grip [hilt] close to the shell; that the grip is strictly embraced by the middle finger and again grasped against the thenar towards the pommel by the ring and little finger. But one must not squeeze the grip except only on the instant that you thrust or that you parry; because the muscles of the thumb, the index finger, and the middle finger will quickly become numb, which is not the case for those that act in the little finger and the ring finger.

There are occasions when it is convenient to release those two fingers to facilitate the execution of certain thrusts. I shall pay attention to warn you when it is needed.

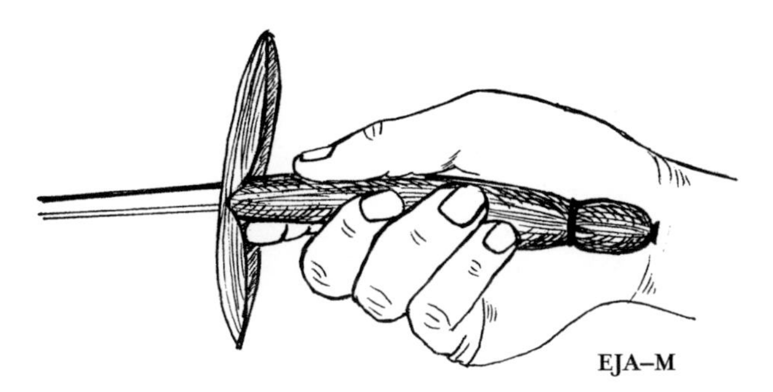

The second manner is identical to the first, except that the index finger is placed directly under the cross and is not extended. This method is described on pp. 1–2 of Charles Besnard's 1653 treatise *Le Maistre d'armes libéral, traittant de la theorie de l'art et exercice de l'espée seule, ou fleuret.*

> *Pour bien se mettre en garde & posture pour faire l'exercice de l'espée seule ou fleuret, il faut premierement mettre l'espée ou fleuret à la main, que le poulce soit pose sur la croisée ou plat de l'espée, & le doigt index soit sous le plat d'icelle en demy rond, & droit sous le poulce, & ferrer fermement la poignée des trois autres doigts, & apres se mettre en garde de cette façon.*

> To well place yourself on guard and posture for the exercise of the single sword or foil,[2] first take the sword or foil in the hand such that the thumb is well placed upon the cross or flat of the sword and the index finger half round directly under the thumb, and hold the grip firmly with the other fingers, and after go on guard in this way.

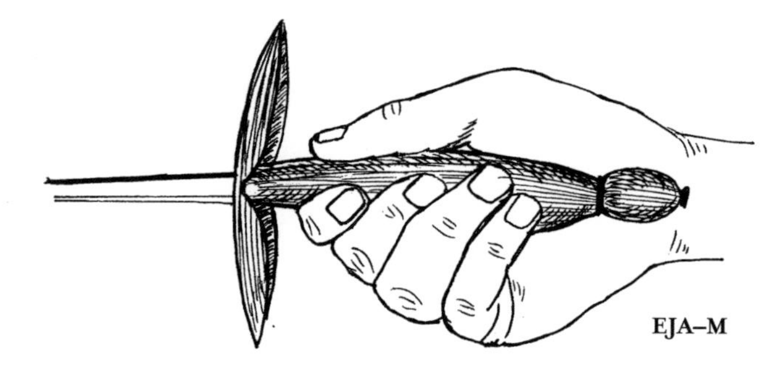

Under either method of holding the weapon, the grip is altered once one begins to move the hand to different positions. For example, in carte the pommel is on the wrist, while in prime, seconde and tierce the pommel is off the wrist. When moving from carte to tierce, the fingers shift slightly in order to swivel the pommel off the wrist.

Angelo does not explicitly describe how he holds the sword, but a close examination of the illustrations shows that the second manner of holding the weapon is most likely. Comparing the plates with the descriptions in the text allows us to double-check the engraver's work. With the exception of a few minor flaws, the illustrations closely demonstrate what is written in the text.

---

[1] While the term *poignée* literally means 'handle', I have translated *poignée* as 'grip' because in this case Danet is talking about the grip of the weapon. Twice I have included the term hilt next to the term grip because Danet does not make himself clear. In these two instances Danet is using the term to mean more than just the grip. As he is talking about holding the *épée* (small sword) and not the foil, the *poignée* must also include the quillon block and ricasso. This is important to understand or one could interpret his description to mean that all five fingers must be compressed together to fit on the grip (see Appendix A, letter D), with no digit on the cross or ricasso. This cannot be the case as he clearly says that the index finger is placed close to the shell.

[2] It is important to understand that the foil was the practice weapon for the small sword; this was made clear as early as 1653 in Besnard's treatise. Mid-seventeenth-century foils were made with a variety of hilts; each of these hilt variants retained some of the elements of sword hilts from the period. The foils in the eighteenth century lost many of these elements; by the end of the second half of the century they consisted simply of the shell, grip, and pommel. It is also important to note that the foil was not held significantly differently than a small sword. Illustrations of some of these different types of foils can be found in Diderot's *Encyclopédie*.

# NOTES

BY MAESTRO JEANNETTE ACOSTA-MARTINEZ

Page xv *To their Royal Highnesses . . . Henry-Frederic.* The original French version published in 1763 dedicates the text to the Princes William Henry and Henry Fredrick. The dedication page in the English version of 1787 changes the name William Henry to the Duke of Gloucester, as this became his title in 1764. The two princes were the sons of Fredrick Louis Prince of Wales. It was their elder brother George William Fredrick who succeeded George II to become George III, King of England.

*teaching your . . . Art of Fencing.* The Dowager Princess of Wales appointed Domenico Angelo to teach fencing and riding to her sons. She also provided Leicester Fields as a place for their instruction. In 1759 Angelo moved there with his family (J. D. Aylward, *The House of Angelo*, London, 1953).

xviii *attitudes.* The term, used on the title page and repeated in this section, specifically refers to the various positions and postures that the body should assume in the art of fencing. For example, the position of the body while on guard, in a lunge, when making a pass on the attack, etc.

xix *compilers of the French Encyclopedia.* Henry Angelo is referring to Denis Diderot and Jean le Rond d'Alembert, the compilers of the *Encyclopédie, ou dictionnaire raisonné des sciences, des métiers et des arts.* The *Encyclopédie* was comprised of seventeen volumes of text and eleven volumes of plates, and was published between 1751 and 1772. Domenico Angelo's text was selected for the volume on fencing. According to Carl A. Thimm in his *A Complete Bibliography of Fencing and Duelling,* published in 1896, the 1787 English version was 'translated into French, and reproduced together with the plates, under the heading "Escrime", by Diderot and d'Alembert in their *Encyclopédie*'. However, this is erroneous. The version of 'Escrime' that Thimm refers to is an exact duplication of the original French text, written by Angelo in 1763. The only changes are the deletion of the dedication and a few sentences written in reference to the British nation. Also, in the editing process pronouns were changed from Angelo's original 'I' to 'we', 'one' or 'it', in addition to a few other grammatical changes.

1–2 *The Method of Mounting a Sword*

Typically in the eighteenth century one went to a furbisher to buy a small sword. One would select the blade and the mountings, and the furbisher would assemble them. Diderot's *Encyclopédie* has a section titled 'Fourbisseur' which contains plates of weapons, weapon parts, tools and an illustration of a sword cutlery shop.

1 *tongue.* The tang of the blade: see Appendix A.

§

*broad-swords.* 'A general term applied from the seventeenth century to heavy military swords with a large double-edge blade, designed mainly for cutting, and either a basket hilt or a well-developed shell guard.' Leonid Tarassuk and Claude Blair (eds), *The Complete Encyclopedia of Arms and Weapons*, London, 1982, p. 105.

§

*half-spadoons.* This is a translation of *demi-espadons*, which is the term used in the French text. This refers to the spadroon, which, according to Tarassuk and Blair in their *Encyclopedia*, was the 'name given in the eighteenth and nineteenth centuries to a light, simplified version of the Broadsword. It had a fairly narrow cut-and-thrust blade and closed hilt with a knuckle guard and rear quillon.' I have found no evidence of half-spadoon being a reference to any other type of sword.

2 *heel of the fort.* What is commonly called the shoulders of the blade: see Appendix A.

§

*which should . . . in carte.* The meaning is clearer if we look at the original French text: '. . . *laquelle doit baisser un peu sur les doigts de la main, & le corps de la garde doit être tourné un peu en quarte'* (. . . which has to fall a little on the fingers of the hand, and the mounting must be a little turned to carte). It was common practice in the eighteenth century to cant the blade. Canting is a bending of the blade where the heel meets the tang. When the small sword is held in the hand in the position of carte, thumb on top, the cant is slightly downward and to the left.

§

*Some . . . hollow blades.* Flat blades are double-edged blades of a variety of cross-section. Hollow blades are triangular in cross-section.

3 *that there be no flaws . . .cross the blade.* Most masters included instruction on how to choose a blade. André Wernesson, Sieur de Liancour, in his 1686 treatise *L'Exercise de l'épée seule dans sa perfection,* states: 'To recognise quality it is good to examine thoroughly the blade from the point to the tang . . . To check for flaw points do the following. The flaws are like small holes. Some go across, others lengthwise. The latter are not so dangerous.' These types of flaws would be visible to the eye when examining the blade.

4 *You must stand . . . your legs.* As can be seen in plate 1, the knees are straight; the left heel is just slightly in front of the right toe. In order to have the point of your right foot in line with your knee, you must

have your weight centred between your left heel and right toe.

§

*the hook of your scabbard.* A hook or sometimes a button-like protrusion on the upper furniture of the scabbard. However, most surviving scabbards appear to have been mounted with two rings, one at the upper part of the scabbard and the other halfway down: see Appendix A.

5 *hold tight and firm.* Angelo does not give us any more information on how to hold the sword than this. See Appendix B for the manner in which to grip the weapon.

§

*presenting the point . . . adversary.* This is a description of how the point and arm make a straight line facing the adversary.

6 *you must observe . . . flexible.* As in most treatises the descriptions are for right-hand fencers. The left knee when bent does not go past the toes of the left leg. This is critical for ease of movement. The right leg is held almost straight with just a slight bend of the knee; if one bends the left knee too much it will result in the body inclining too far back and to the left, throwing the balance off.

§

*longe.* An eighteenth-century spelling of 'lunge'.

§

*low parts.* Described in the original French simply as 'below the weapon'.

7 *The carte . . . outside one.* When holding the small sword with your fingernails up and the thumb directly on top, in other words in the carte position, the inside edge faces to your left and is on the same side as the knuckle guard. The outside edge faces to your right. When your hand is in tierce the inside edge is to

your right and the outside edge is to your left.

§

*The prime . . . swords.* The original French says '*La prime doit être tirée au-dedans des armes*' (Prime should be thrust within the swords). This means that prime is thrust to the inside of the adversary's sword.

§

*The carte over the arm.* Thrust with hand position in carte and the blade edges level. However, the thrust covers the outside as when thrusting in tierce.

§

*The low carte.* The same as carte except that the elevation of the point changes – the thrust is now directed below the wrist.

§

*The seconde.* Thrust under the adversary's wrist, covering your outside and lower parts.

§

*The quinte.* The same hand position as carte over the arm except that the thrust is directed below the adversary's wrist.

8 *To get . . . too near.* When the adversary is too far from your point for you to be able to hit him on your attack you must advance to 'within distance'. Within distance is the distance that you must be in to be able to hit your adversary on your lunge. If you are in danger of being hit you may break distance by retiring backwards. This is to be 'without distance'.

9 *double advance.* The French text simply says '*On ferre aussi la mesure*' (We also get within distance by). This type of footwork is designed to facilitate gaining more distance on the step. On a normal advance the right foot moves forward roughly the length of your own foot length. The left foot then immediately moves

forward the same distance, such that the distance between the feet is identical before and after the advance. On the 'double advance' one moves the rear foot first by bringing it up as far as the right toe. Then the right foot moves forward roughly two foot lengths. Therefore, on this type of advance you actually advance twice as far as the normal advance – hence the English name. This also applies to stepping back from your adversary.

*jump back about two feet.* Some masters taught a type of footwork that allowed for leaping back to get out of measure quickly. This leap or jump back must be done in such a fashion that the fencer travels directly backwards without moving his body upwards, as one would typically visualise in reference to jumping.

9–10 *Position for the Guard in Tierce*

This section is a little confusing because one could get the impression that you begin by engaging in tierce. It becomes clear if one refers to the French text, as it says '*il faut exécuter des dégagemens qui se sont en changeant la position de la main & de la pointe, au-dedans ou au dehors des armes*' (to perform the disengagements, which are by changing the position of the hand and the point, inside or outside the weapon). So Angelo is not telling us to begin by engaging in tierce but rather he is describing how the change of engagement is done. The contact on these engagements is made with your foible to his foible.

What Angelo is explicating is a drill to facilitate learning how to engage, disengage and make a change of engagements. The drill begins from the engagement in carte.

9–10 *so that . . . carte to tierce.* Thus being engaged in carte to make a change of engagement it is important to disengage by turning the wrist as you drop the point and move it from the inside to the outside where you engage the adversary's blade again in the position of tierce.

10 *in a firm position.* This is referring to the body remaining in a stationary position. In this drill only the hand is moving from carte to tierce and from tierce to carte.

§

*till your . . . on your guard.* This is a new element added to the drill. The adversary now steps back (keep in mind that he will maintain blade contact until he has finished his step back), you disengage as he does so and then advance with your arm extended, your blade against his, with your point aimed at his chest. The posture of your body does not change on the advance. The French text says '*sans altérer nullement la position de la garde*' (without altering at all the position of the guard). This drill is done both in carte and in tierce.

§

*you must retire . . . lower parts.* Now Angelo has added a third element to this drill. This time your partner will do the engagement exercise and you will step back. He will disengage and advance with his arm extended, his blade against yours. Every time he does so, you turn your wrist to the position in which you are engaged. For example, you step back with your hand in the tierce position. He advances with his arm extended in his carte position. You turn your hand to carte. As you turn your hand your blade does not lose contact with your partner's blade while at the same time it deviates his point. That is the parry. On the parry you oppose his blade only enough to sufficiently cover the inside line, which is the side he is threatening. At the same time you must be observant not to parry so wide as to uncover your outside or lower parts.

11 *oppose.* In small-sword theory, oppose means to apply pressure with your blade on your adversary's blade in order to deviate it. This is done both offensively and defensively.

§

*being thus situated . . . the left.* This is the description of the lunge. The English version is a bit confusing, as it sounds as if you have two options – one to step in quickly on the attack, the other to lunge on the attack. The French text clearly says '*avancer le pié droit de deux semelles de plus que la garde pour s'étendre, le talon & le genou*' (advance the right foot two feet more than the guard to extend, the heel and the knee). Angelo is simply telling us that the right foot must advance on the lunge. Also the 'left heel' in this case is an erratum as he is referring to the right heel.

§

*the left foot . . . ground.* On the lunge the left foot must stay perfectly flat on the ground. Some masters advocate turning the left foot on its edge. See Monsieur L'Abbat's *The Art of Fencing, or The Use of the Small Sword*, translation by Andrew Mahon, Dublin, 1734, p. 10. The turning of the foot is sometimes advocated as a way of lunging further. It is more often explained as a tactical consideration when lunging on terrain. Turning on the edge was believed to allow you to dig into the ground to prevent you from slipping or falling on the lunge.

12 *This position . . . in carte.* In the French text this line is a separate sentence. The line contains another erratum, as it should say the left hand. Angelo is telling us that every time the hand is turned to carte on the thrust one must observe the left hand should be in the same orientation. That is to say, that the left hand should be palm facing up. This is an important distinction because he will go on to tell us that every time the thrust is made in tierce the left hand should be palm down.

§

*the sword in a strait line.* On the recovery from a lunge the last thing back is the arm. That is to say, the arm remains extended posing a threat to the adversary until the recovery is completed. Bear in mind that this is the way to practise the lunge in carte.

13 *the left . . . the same.* This section is potentially confusing. When it says, 'Observe, that at all times when the right arm is turned with your nails down, that the left should be the same,' it would seem to imply that the fingernails of the left hand should be down. Yet, Angelo has already told us that the nails of the left hand are upward in the tierce thrust. It is clearer to consider the orientation of the palm. In a tierce thrust, both palms face down.

16 *but strait . . . foot outward.* The French text gives a more accurate description: '*mais encore porter le poignet & le pié droit à un pié en dehors de la ligne directe*' (but still carry the wrist and the right foot to the outside of the direct line). In this case you lunge so that your right foot is almost one foot to the outside of the direct line. The outside in this case refers to the adversary's outside.

Angelo tells us to carry our right foot to the outside of the 'direct line', but he does not explain what he means by direct line. Clarification can be found in Monsieur L'Abbat's *The Art of Fencing*:

> The line must be taken from the hindmost part of the right heel to the left heel near the ankle. The point of the right foot must be opposite to the adversary's, turning out the point of the left foot, and bending the knee over the point of the same foot, keeping the right knee a little bent, that it may have freedom of motion.

What is significant here is that both fencers must stand upon an imaginary line. That is why L'Abbat tells us that the point of the right foot must be opposite the adversary's right foot. He further explains this when discussing the lunge:

> The foot should go out strait; in order to preserve the strength and swiftness of the thrust, it must have its proper line and distance. The line must be taken from the inside of the left heel to the point of the adversary's right foot; if it turn inward or outward, the button will not go so far, the strait line being the shortest; besides, the body would be uncovered, for by carrying the foot inward, the flank is exposed, and by carrying it outwards the front of the body, the body is thereby weakened.

16 *the sword . . . in carte.* In order to do this thrust, the adversary must have his arm extended to your inside. If there is no blade contact then you engage his blade in carte. If he is extended with opposition then all you have to do is turn your hand to carte.

19 *carrying . . . the right.* The English text is vague, but the French text is very clear as it specifically tells us to carry the left leg back.

20 *Being . . . tierce guard.* Having carried the left leg back the two fencers cannot be 'engaged' in tierce (see previous note). The French text makes it clear that Angelo is not referring to an engagement (blade-to-blade contact) of tierce but rather to the fencers having assumed the position of tierce guard.

22 *elastic disengagement and disposition.* This is not in the French text. It is referring to suppleness of the wrist, a key element in the French system of small sword.

§

*the motion . . . whole frame.* Again, looking at the French text this becomes more clear, as it says '*faire mouvoir les jointures de toutes les parties du corps comme des ressorts*' (move the joints of all the parts of the body as springs). He is referring to the suppleness and light pliancy of the body that allows for fluidity of motion.

§

*return, or reposte.* Here, the terms are synonymous. The French text simply says 'reposte', as Angelo makes no distinction between the terms. Some masters, however, do make a distinction: the 'return' refers to the response after the parry while the adversary is recovering. The 'reposte' refers to an immediate response after the parry while the adversary is still in the lunge or pass. There are also variations in the timing of these actions, which can further complicate the distinction. In addition, the manner in which the parry is executed affects the timing.

§

*he must carry . . . the right.* This is a pedagogical method for the student to practise ease of movement while in correct balance. The student recovers forward or backward by bringing his feet close together. If the balance is not correct the student may lose his equilibrium on the recovery.

§

*plastroon.* A 'plastron'. When practising thrusts, the fencer usually thrusts at the master, who wears a plastron for that purpose. The plastron is a padded chest protector that the master wears to protect himself from any injury that may result from repeated thrusts to his chest.

23

*to answer . . . plate XIV.* The French text says '*comme on peut le voir à la quatorzieme Planche dans la cinquieme position du salut*' (as one can see in the fifteenth plate in the fifth position of the salute). This clarifies 'to answer', as it simply means the same as the fifth position of the salute.

§

*carry the point . . . right toe.* In this section Angelo is introducing new elements of footwork to give suppleness and strength to the legs and ensure balance. On the lunge in tierce, instead of recovering backward one recovers forward by bringing the left foot forward in front of the right foot, keeping the knees bent. From this position one can go on guard by bringing the right foot forward or bringing the left foot backward.

§

*The master . . . and legs.* When the student is on the lunge the master will hold the scholar's right wrist with his left hand to adjust the student's position and balance. This will assist the student on his recovery, as he will be learning to recover from a properly positioned and balanced lunge.

§

*It is very necessary . . . foil.* The master will allow the student to practise his thrust against him on the plastron. He will also deliberately step back, withdrawing the plastron from time to time. This is important so that the student does not become accustomed to lunging at a target and using the target to stop his forward momentum. By stepping back on some thrusts, the master forces the student to work on his balance and be in control of his movements.

26

*You should apply . . . the point.* This is another erratum, as the French text says '*On doit donc s'appliquer à bien former ses parades, en tenant ferme son épée depuis la garde jusqu'à la pointe*' (We thus have to apply ourselves to form well our parries, by holding firm our sword from the guard up to the point). Therefore, Angelo is not telling us to hold the sword 'light', but rather that on the parry it is important to hold the sword firmly in your hand, controlling it from the guard to the point. In this fashion your sword will dominate the opponent's. Guillaume Danet says '*mais il ne faut ferrer la poignée que dans l'instant seulement que vous tirez, ou que vous parez*' (but one must not squeeze the grip except only on the instant that you thrust or that you parry), *L'Art des armes*, 1767, p. 3. Also, on p. 21, Angelo says: 'In the art of fencing, much depends on a quickness of sight, agility in the wrist, a staunchness in the parades . . .'

§

*singled out.* This simply means being well profiled.

27 *Of the Outside Parade, Called Tierce, and the Tierce Thrust, Called the Outside Thrust*

Angelo is describing two types of parries, both of which are done with the inside edge with the hand in tierce.

The first is done with the straight arm. The French text says '*sans le déranger de la position de tierce*' (without leaving the position of tierce). The English text says 'without leaving the strait line'. Both are referring to the hand staying in the line of tierce. The second parry of tierce is done with a bent elbow followed by a riposte in seconde.

28 *Of the Outward Thrust, with the Nails Upward, Commonly Called the Feather Parade, against the Outward Thrust, Nails Upward, Called the Carte over the Arm*

Angelo describes two types of parries. The first is done with a straight arm. The second parry is done with a bent elbow and the point raised, this position allows you to then do a glizade as described on p. 25.

30 *The reversing . . . to feeble.* When the adversary binds your blade in a flanconade, you simply turn your hand from carte to seconde. The flanconade begins from the inside of your blade and ends on the outside of it. You cavé when he is on the outside of your blade. Seconde, being a strong parry position, will easily deviate his point away from your flank. Also, since your point will be in a direct angle to his body and your arm is extended, he will end up running on to your point.

31 *gathering his blade in carte.* This parry, called a bind, is not the act of transporting the blade; it is in fact what became know as a ceding parry in the nineteenth century. The French text says '*Il faut céder la pointe*' (It is necessary to yield the point). When the adversary binds your blade you permit your foible to be taken by the force and leverage of the adversary's binding action, allowing your point to pass under his wrist. Do not detach your blade at any time; as the thrust is directed towards your flank, catch his foible with the forte of your blade as you pick up your point. Your hand will be in carte. Angelo defines the 'bind' on p. 53.

*Of the Parade . . . Called Seconde.* The translator, or perhaps Angelo himself, added this title to the English version, no doubt to cater to the sensibilities of his English patrons. The French text titles this section '*De la Parade de prime sur le coup de seconde*' (Of the Parry of Prime on the Thrust of Seconde).

32 *hang down . . . the same manner.* This refers to an opposition with the left hand. The left hand is not used to parry, only to keep the adversary's blade in check after your blade has already parried it. The parry is done with the sword and the left hand is brought in to secure the blade in that line on your riposte. In this manner the adversary cannot place the point back in line.

33 *The thrust . . . at his flank.* The quinte thrust and parry are on the same side as tierce and carte over the arm. However, the wrist is held in a high carte with the point low.

35 *baffle.* This word is not in the French text; it is just another way of saying that, if you know how to defend with a well-executed parry, you will frustrate the adversary and be able to riposte quickly.

37 *If the adversary . . . demi-circle.* On the recovery it is important to protect yourself from a counter attack, so Angelo advocates recovery by taking the adversary's blade to another line. In this section he advocates thrusting in low carte and recovering with the wrist in tierce. One can also recover in demi-circle to cover the line the adversary is in.

40 *forming an angle . . . his body.* Once you have turned your wrist in tierce make sure that an angle is created by having your hand to your outside while directing your point to the adversary's flank.

ʃ

*circle.* Angelo means the half-circle parade.

41 *support . . . to feeble.* As Angelo explained before, when you parry you must hold the sword firm from the guard up to the point.

43–4 *Method of Thrusting and Parrying Tierce and Carte, Called Thrusting at the Wall*

Angelo has explained to us how the student must thrust at the plastron and the reason for this. Now he is giving us an exercise that is designed for two students to work on perfecting the accuracy of their thrust, their awareness of proper lunge distance to various persons of varying size, and the swiftness of their thrust and parry. Keep in mind that thrusting-at-the-wall drills are purely pedagogical exercises.

43–4 *open with . . . his longe.* To begin with the student who remains stationary brings his point off line to the right to allow his adversary to take the proper distance for his lunge. Keep in mind that the person lunging sets the distance for the drill, so he may appear to be too far or too near to the person who is parrying at the wall. It is essential that the student working on the parries does not move his feet at all. This allows him to develop confidence and exactness in the placement of his parries, as he must rely only on his hand.

44 *button*. A covering that is placed over the nail-shaped tip of the foil.

45 *by engaging . . . seeking his blade*. This is the first 'thrusting at the wall' exercise. Keep in mind that this is an artificial drill, intended to help the student develop a supple wrist and correct finger movement on the disengagements; this exercise is *not* intended to teach combative technique.

You are asked to engage the fort of the adversary's blade with your foible, requiring that your arm be extended. It should be noted that this is not a realistic action; one would never engage one's foible to the opponent's forte, as he would have all the mechanical advantage. Your adversary will move your foil out of line (i.e. parry) with a small movement of his wrist. You must be sensitive to this small movement or pressure on your foible, because you will disengage immediately upon sensing it, thereby freeing your point, and immediately thrust. This is why Angelo recommends 'holding a loose point'. On the thrust do not seek the adversary's blade. Although the English text says 'seeking his blade', this is a translation error. The French text clearly says '*tirer droit au corps de l'adversaire sans chercher nullement sa lame*' (thrust straight to the body of the opponent without looking at all for his blade).

§

*by engaging . . . his blade*. The second way of thrusting at the wall begins with a normal engagement of foible to foible.

§

*by delivering . . . nor disengage*. The third way of thrusting at the wall is by delivering straight thrusts without any engagements or disengagements of any type. The adversary will give openings in either tierce or carte always being ready to parry once you thrust.

47

*treble.* In this instance, 'treble' means 'triple'.

§

*To feint . . . strait at him.* By pressing on the adversary's blade with your own, you bring his point off line, thus forcing him to disengage to bring his point back in line. You are actively seeking his disengage in order to take the opportunity to feint. At the moment he slips or disengages you will feint and when he attempts to parry your feint you will thrust to the open line.

48

*The surest defence . . . and idea.* Angelo advocates using the counter-disengage or the circle parries to defend against feints. See the sections discussing the parade by counter-disengage and the half-circle parade on pp. 41–3.

§

*all these attempts, which are false.* Angelo says that these attempts 'are false' because all motions that do not adhere to the laws of fencing are useless against a skilful adversary. In order to force the adversary to defend, one most either pose a real, or present a believable perceived, threat. If the adversary does not feel threatened, he need not answer, and the feint will be ineffectual.

49

*who seeks . . . wrist closely.* The French text is clearer, as it says '*qui recherche l'épée de son adversaire avec le mouvement seul du poignet*' (who looks for the sword of his opponent with only the movement of the wrist). Angelo has already discussed the importance of closing the line on the parry by making small movements of the wrist: see p. 10.

§

*drawing and changing . . . thrust out.* Angelo is referring to the practice of drawing the arm back to disengage when the adversary attempts to parry the feint. He tells us that this is dangerous because it takes too long, and the adversary may attack in the moment that you draw your arm back.

50 *Glizades.* The glizade is described on p. 25.

§

*beat.* A 'beat' is an abrupt strike on the adversary's blade with the purpose of moving it out of line. It is executed by striking your forte to his foible, edge to edge. Sometimes it also causes the adversary to lose control momentarily of his blade by jarring his grip.

§

*carte over the arm.* The glizade is done on the carte over the arm. As you glide down his blade the adversary will parry with opposition to move your point off line; at that moment you disengage and thrust in carte.

51 *you must disengage . . . thrust out.* The act of gliding down his blade along with the appel of the foot will cause the adversary to oppose your foible to move it off line. At that moment you disengage and thrust carte over the arm. 'Slip' refers to the disengagement.

§

*If you are . . . tierce side.* This is an erratum in the English translation. The French text says '*si on est engagé en quarte*' (if one is engaged in carte).

§

*parry.* The French text makes it clear that it is the adversary who is executing the parry, either in tierce or in carte over the arm.

*the moment . . . thrust seconde.* On your glizade in tierce when the adversary feels your blade he will make a simple parry of tierce or carte over the arm. You will then disengage and make a thrust in seconde.

52 *redouble.* In this case to 'redouble' means to recover and instantly thrust again before the adversary has a chance to riposte. This can be done in tierce, if you have recovered in tierce, or you can disengage and go to another line on the thrust.

53 *Binding and Crossing.* This would suggest that 'binding' and 'crossing' are two separate actions. The French title simply reads '*Le Croisé d'épée*' (The Crossing of the Blade).

*you should incline . . . left side.* This is misleading as one could take this to mean that you must incline your body backwards or directly to the left. The French text is clearer as I states '*il faut poser bien le corps sur la partie gauche*' (it is necessary to put well the body on the left part). This is in reference to being well positioned while on guard. This is very important because if you bring your body at all forward you expose yourself to a thrust if your adversary should disengage on your attempt to bind his blade.

61 *Of the Definition of the Wrist, after the Thrust Made in Carte*

In the title of this section, the meaning of 'Definition of the Wrist' is initially unclear. The French text reads '*Du Coup de reprise de la main après avoir tiré le coup de quarte*' (Of the Resumption of the Thrust of the Hand after the Thrust of Carte). The word '*reprise*' means to resume or repeat. The meaning becomes clearer upon reading, however – the sequence described is a false attack designed to draw the parry. On the recovery from the attack, one does not recover completely but only seems to be recovering by bringing back the lead foot about a foot distance. Once the adversary begins to come forward you intercept him by swiftly turning your wrist from carte to tierce or from tierce to carte over the arm.

62 *pass.* The 'pass' is another type of footwork used on the attack. This is the first time Angelo introduces this footwork.

62–3 *then without altering . . . left leg.* In the guard position, the feet are approximately two foot lengths apart. In the lunge, Angelo instructs us to bring the right foot forward another two foot lengths. On the pass he tells us to bring our left foot forward of the right foot about two foot lengths. The French text uses the term '*semelles*', which means 'soles'. Therefore the distance one moves forward on the pass should be the same distance as that of your lunge.

73 *tierce edge.* This refers to the inside edge of the sword. The inside edge faces to your right when you are in the tierce guard.

76 *force or bear his wrist upwards.* The French text tells us to '*forcer au haut son poignet; dans le même instant saisir promtement avec la main gauche la garde de son épée*' (force high his wrist: in the same moment promptly seize with the left hand the guard/shell of his sword).

§

*throw his arm . . . the right.* The French text make it clear that you are moving his arm out to his right: '*écarter son bras sur sa droit*' (push aside his arm on his right).

77 *right arm.* This is another erratum, as the French text instructs us to bring in the left arm.

§

*by drawing in . . . left arm.* The text is unclear when it says 'by this means, by drawing in your body and your left arm'. The French text clarifies this by stating '*par ce moyen en retirant le corps & raccourcissant le bras gauche, il sera forcé de céder son épée*' (by this means, in removing the body and shortening the left arm, he will be forced to give up his sword). Angelo does not give us a complete description of this disarm, but he does give us very detailed illustrations which allow us to see the elements that have been left out. We are told to advance about a half-foot on the parry of prime; the illustration (plate 38) shows that the advance is done by passing the left foot forward of the right. This is an important element of this disarm, as it brings the left side further forward, allowing the left hand to be brought over the adversary's fort and then under his wrist to seize the shell of his sword. Although the illustration appears to show the adversary's wrist being grabbed, we can be confident that it is actually the shell because, at the end of the section on disarms, Angelo explicitly tells us not to seize the adversary's arm or blade with the left hand.

Although the element of grabbing the shell of the adversary's sword is not mentioned in this disarm, it is crucial since it prevents the adversary from retiring. The grip on the shell is sometimes enough to make the adversary surrender the weapon. Alternatively, since you are holding on to his weapon and he cannot retire, the adversary may recover forward to counter the attempted disarm: see plate 39. As he does this, he will bring his left hand forward in an attempt to prevent you from drawing your hand back preparatory to presenting your point. Therefore, when the adversary comes forward it is important that you do three things: draw your left arm into your chest, thus locking his weapon under your arm; turn very slightly from the waist to create an angle that makes it even more difficult for the adversary to pull his weapon out; and bring the right leg back while bending the left knee. This puts you in a better position from which to present the point.

81 *the right handed . . . thrusts without.* When a right-hander is on guard in carte, his adversary, if right handed, will also be in carte. The same for tierce – when one is in tierce the other will be in tierce. With a left-hander, when he is on guard in carte the right-handed adversary should be on guard in tierce. When he is in tierce the adversary should be in carte. Therefore, the carte thrust from the right-hander will be to the outside of the adversary's weapon and the tierce thrust will be to the inside of the adversary's weapon. The left-hander when thrusting in carte will thrust to the outside of the adversary's weapon, and when in tierce to the inside of the adversary's weapon. If two left-handers face each other it is the same as two right-handers facing each other.

82 *the German Guard.* Thimm, *A Complete Bibliography*, catalogues a number of eighteenth-century treatises on the German School.

83 *you should parry . . . the right.* On the parry, one steps about a foot to the right to facilitate the riposte. This was discussed on the section on parries: see p. 32.

84 *the Italian Guard.* Bibliographical references to Italian treatises on fencing theory and practice written in the eighteenth century are scarce. To better understand the Italian guard as Angelo describes it, refer to:
*La Spada Maestra*, Bondi Di Mazo, 1696
*La Vera Scherma Napolitana*, Nicola Terracusa e Ventura, 1725
*Elementi della Scherma*, Marco Marcello Vandoni, 1750
*Regionamenti Accademici Intorno all'Arte della Scherma*, Alessandro Di Marco, 1758
*La Scienza della Scherma*..., Rosaroll Scorza and Pietro Grisetti, 1803.

92 *the Spanish Guard.* For the Spanish school in the eighteenth century, see Don Francisco Lorenz de Rada, *Nobleza de la Spada*, Madrid, 1705.

*longe.* For a description and illustration of this lunge, see Rada, *Nobleza*, pp. 394–5.

93 *Their swords . . . to point.* This is erroneous. Rada, *Nobleza*, discusses the two methods of determining the proper measurement of the sword on pp. 503–6:

1. Extending the left arm horizontally and aligning the sword by placing the pommel in line with the right shoulder with the point of the sword in line with the extremity of the middle finger determine the proper length for the sword.

2. The proper length of the sword is determined by standing straight holding the sword vertically with the point on the floor between one's feet, the quillons crossing the naval.

99 *redouble it . . . the lanthorn.* One delivers a cut which targets the hand holding the lantern. This cut is from the elbow, with the weapon hand in tierce, and uses the outside edge of the blade.

*to the right.* The French text stipulates to the adversary's right.

100 *those keep a straight arm.* This could lead the reader to believe that the arm is simply extended and thus straight. The French text says '*Ils tiennent le bras roide & tendu*' (They hold the arm rigid and tense).

101 *appuy.* This means 'support'.

105 *counter point.* In the original French, the term reads '*Faire la contre pointe*'. A broadswordsman may feint with a cut then turn it into a thrust, or on his parry he may riposte by cut or thrust.

# Select Bibliography


Angelo, Domenico, *L'Ecole des armes, avec l'explication général des principales attitudes et positions concernant l'escrime*, London, 1763

——— *L'Ecole des armes, avec l'explication général des principales attitudes et positions concernant l'escrime*, London, 1765

——— *L'Ecole des armes, avec l'explication général des principales attitudes et positions concernant l'escrime*, London, 1767

——— *The School of Fencing, with a General Explanation of the Principal Attitudes and Positions Peculiar to the Art*, London, 1787

——— *The School of Fencing, with a General Explanation of the Principal Attitudes and Positions Peculiar to the Art*, New York, 1971

Angelo, Henry, *Reminiscences of Henry Angelo*, London, 1830

Aylward, J. D., *The House of Angelo*, London, 1953

——— *The Small-Sword in England*, London, 1960

Besnard, Charles, *Le Maistre d'armes libéral, traittant de la théorie de l'art et exercice de l'espée seule, ou fleuret, et de tout ce qui s'y peut faire et pratiquer de plus subtil, avec les principales figures et postures en taille douce; contenant en outré plusieurs moralitez sur ce sujet*, Rennes, 1653

Blackwell, Edward, *A Compleat System of Fencing, or The Art of Defense, in the Use of the Small-Sword*, Williamsburg, 1734

Blackwell, Henry, *The Gentleman's Tutor for the Small Sword, or The Compleat English Fencing Master*, London, 1730

Blair, Claude, et al., *Studies in European Arms and Armor: The C. Otto von Kienbusch Collection in the Philadelphia Museum of Art*, Philadelphia, 1992

Danet, Guillaume (Syndic-garde de la compagnie des maîtres d'armes de Paris), *L'Art des armes*, Paris, 1766–7

Di Mazo, Bondi, *La Spada Maestra*, Venice, 1696

Diderot, Denis, and Jean le Rond d'Alembert, *Encyclopédie, ou dictionnaire raisonné des sciences, des métiers et des arts*, 28 vols, Paris, 1751–72

Girard, P. J. F., *Traité des armes*, La Haye, 1740

Gordine, Gerard (Capitaine et maitre en fait d'armes), *Principes et quintessence des armes*, Liege, 1754

Hope, William, *The Scots Fencing Master*, Edinburgh, 1687

L' Abbat, *The Art of Fencing, or The Use of the Small Sword*, trans. Andrew Mahon, Dublin, 1734

La Boëssière, *Observations sur le traité de l'art des armes, pour server de défense à la verité des principes enseignés par les maîtres d'armes de Paris, par M. maître d'armes des académies du Roi, au nom de sa compagnie*, Paris, 1766

La Touche, Philibert de, *Les Vrays Principes de l'espée seule,* Paris, 1670

Le Perche, Jean-Baptiste, *L'Exercise des armes, ou le maniement du fleuret,* Paris, 1676

Liancour, André Wernesson, Sieur de, *L'Exercise de l'épée seule dans sa perfection,* Paris, 1686

Lonnergan, A., *The Fencers Guide,* London, 1771

McArthur, J., *The Army and Navy Gentleman's Companion,* London, 1780

McBane, Donald, *Expert Sword-Man's Companion,* Glasgow, 1728

Massuet, P., *La Science des personnes de cour, d'épée et de robe. Commensé par de Chevigny, continué par de Limiers, revue, corrigé et augmenté par P. Massuet,* Amsterdam, 1752

Menessiez, *Mémoire pour le sieur Menessiez, maître en fait d'armes, et maître des pages de M. le comte de Clermont. Contre la Communauté des maîtres en fait d'armes,* Paris, 1763

Norman, A. V. B., *The Rapier and Small-Sword, 1460–1820,* London, 1980

North, Anthony, *European Swords,* London, 1982

Oakeshott, Ewart, *European Weapons and Armor,* Guildford, 1980

Olivier, J., *Fencing Familiarized, or A New Treatise on the Art of Sword Play,* London, 1771

O'Sullivan, Daniel (Maitre en fait d'armes des académies du Roi), *L'Escrime pratique, ou principes de la science des armes,* Paris, 1765

Rada, Don Francisco Lorenz de, *Nobleza de la Spada,* Madrid, 1705

Scorza, Rosaroll and Pietro Grisetti, *La Scienza della Scherma,* Milan, 1803

——— *La Scienza della Scherma,* 1871

Tarassuk, Leonid, and Claude Blair, *The Complete Encyclopedia of Arms and Weapons,* New York, 1982

Thimm, Carl A., *A Complete Bibliography of Fencing and Duelling,* London, 1896

——— *A Complete Bibliography of Fencing and Duelling,* New York, 1968

Weischner, C. F., *Exercices dans les salles d'armes,* Weimar, 1752